Renewable Energy

Volumes 1, 2, and 3

Synthesis Lectures on
Renewable Energy Technologies

Editor
Richard A. Dunlap, *Dalhousie University*

Renewable Energy: Volume 1: Requirements and Sources; Volume 2: Mechanical and Thermal Energy Storage Methods; Volume 3: Electrical, Magnetic, and Chemical Energy Storage Methods
Richard A. Dunlap

Combined Edition, Volumes 1–3
ISBN: 978-3-031-01393-5 print
ISBN: 978-3-031-02521-1 ebook
ISBN: 978-3-031-00329-5 hardcover

DOI 10.1007/978-3-031-02521-1

A Publication in the Springer series
SYNTHESIS LECTURES ON RENEWABLE ENERGY TECHNOLOGIES
Lecture #5, 6, 7

Series Editor: Richard A. Dunlap, Dalhousie University

Series ISSN 1940-851X Print 1942-4361 Electronic

Acknowledgments

I am grateful to Nicki Dennis and Joel Claypool for their support and encouragement during the development of this book and to Karen Donnison for her work as Permissions Editor. I am also grateful to Melanie Carlson and Deb Gabriel at Morgan & Claypool for their work on the production of the book.

Renewable Energy

Volume 1: Requirements and Sources

Richard A. Dunlap
Dalhousie University

SYNTHESIS LECTURES ON RENEWABLE ENERGY TECHNOLOGIES #5

ABSTRACT

This book reviews the past and present energy use of society and its future needs. A breakdown of current energy sources shows that approximately 80% of the world's primary energy comes from fossil fuels. The book provides an assessment of the needs to change the way in which energy is produced and utilized. The reasons for change fall into two broad categories; diminishing resources and environmental impact. The Hubbert model is described as a means of projecting availability of fossil fuel energy resources in the future. The environmental impact of fossil fuel use is described, with particular emphasis on global climate change. The major options for carbon-free energy are presented. These options include hydroelectric energy and solar energy for both thermal applications and the production of electricity, wind energy, and biofuels. Renewable energy options that range from residential wind turbines and photovoltaics for electricity and solar thermal heating systems to grid scale facilities, such as off-shore wind farms and hydroelectric installations, are discussed. The production of biofuels as a replacement for fossil fuels used for transportation is also presented. The book also provides evidence for the need to develop energy storage technologies. Energy storage is essential for most forms of renewable energy because the thermal or electrical energy produced by such sources is generally not available when it is needed, nor is it sufficiently portable for transportation applications.

KEYWORDS

renewable energy, sustainability, energy storage technology, climate change, alternative energy

Contents

Preface

Energy is an essential resource of our modern society. Until now, the vast majority of our energy has come from fossil fuels. However, fossil fuels are a resource with a finite lifetime. In just a few decades we have exhausted a substantial fraction of these resources. It has also become apparent that our use of fossil fuel energy has a detrimental, and perhaps even irreversible, impact on our environment. The transition to renewable energy is essential, not only to ensure a continuation of the energy that has become essential to our society, but perhaps more importantly, to avoid continued adverse environmental effects, particularly climate change. Unfortunately, renewable energy resources generally do not provide a constant daily and seasonal supply of energy and they are not, in general, portable enough to provide energy to satisfy our transportation needs. The development of energy storage methods must, therefore, accompany the implementation of renewable energy technologies.

The present volume reviews our historical and current energy use, as well as predictions for future energy needs. The future availability of fossil fuel energy resources, as well as adverse effects of its continued use, are discussed. The possible options for the implementation of renewable energy are discussed along with the details of the need for energy storage methods.

CHAPTER 1

Energy and Society

1.1 INTRODUCTION

Modern society would not exist without the implementation of organized methods for the production and utilization of energy. It is used for heating, cooling, lighting, transportation, and to operate a wide variety of electrical devices that have become essential to our daily lives. Over the years, human energy use has increased. This increase is due to population growth, but also, significantly, as a result of increased industrialization. This chapter reviews our energy use during the past, as well as our present sources of energy and an analysis of future energy needs.

1.2 OUR PAST AND PRESENT ENERGY USE

The earliest humans used only the energy that their bodies produced. This energy came from the food they consumed and amounted to about 2,000 kcal per day. The average power produced during the day was about 100 W per person. Energy use increased when humans began using fire for heating and cooking. Later, humans used energy from domesticated animals for transportation and farming. This was followed by the use of energy from wind and water. In the industrial era, energy use increased substantially with the implementation of steam engines, internal combustion engines and the distribution of electricity. As Figure 1.1 shows, average power consumption per person in the U.S. has leveled off at about 12 kW in recent years. This is, more or less, the result of a trade off between the desire for a more automated life style and the implementation of devices with greater efficiency and energy conservation measures.

It is interesting to look at how the average power consumption of about 12 kW per person is utilized. An individual can account for most of their personal energy use by looking at the ways in which they use different types of energy. This would probably include electricity purchased from a public utility, possibly oil or natural gas for heating and gasoline, or diesel fuel for transportation. Such an exercise will likely yield an average per capita power consumption that is about one third of the value shown in the graph. There are two reasons for this discrepancy.

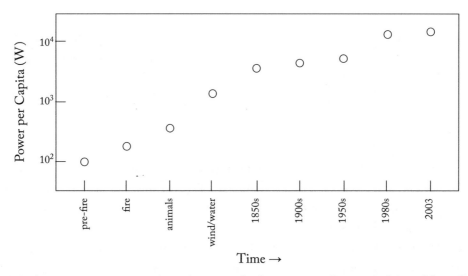

Figure 1.1: Average per capita power consumption by humans as a function of time. Note the logarithmic power scale. Values from 1850 on are typical of the U.S. (data adapted from Aubrecht (2006)).

First, it is important to distinguish between primary energy use and end user energy use. If we calculate the energy content of the gasoline that we put in an automobile, this is very close to the primary energy associated with the oil that came out of a well from which the gasoline was produced. On the other hand, the electrical energy that we use at home, which is measured as the kilowatt hours that we purchase from the public utility, can be quite different from the primary energy that produced that electricity. If, for example, the electricity that we purchase has been produced by burning coal, which is the most common situation in the U.S. and many other parts of the world, then the primary energy, which is the energy content of the coal that is consumed, is quite different than the electrical energy that is supplied to our home. This is because the process of burning coal and producing electricity, which involves using the heat produced by combustion to produce steam, which is then used to drive a turning which turns a generator, is only about 35–40% efficient. So, if we wanted to account for our primary energy use in terms of the electricity we use, we would have to multiply that value by about a factor of about 2.5–3. Of course, if the electricity is produced by some other means, e.g., hydroelectricity, then the factor to account for efficiency will be somewhat different. This correction will probably get us up to about half of the average per capita power consumption.

The second factor that we need to consider is that the average per capita power consumption is a societal average not the average for an individual person. While it is relatively easy to account for energy used in our own home or vehicle, it is more difficult to account for the component of energy that we use, e.g., at work, or that is used in stores or public buildings that we access. There is also energy that is used for the manufacture of goods that we buy and the production of food that we consume. Accounting for this energy use would bring us up to the average power consumption of about 12 kW per person.

It is interesting to contrast energy use in the U.S., which is typical of many industrialized countries, with the situation worldwide. Figure 1.2 shows the total world energy use and the energy use for different regions from 1980–2010. It is clear that world energy use has increased consistently over this period. However, it was suggested previously that the per capita energy consumption in the U.S. has remained relatively constant over this period. This is true for most other highly industrialized countries. Figure 1.2 shows that the total energy use has increased only slightly over this period for the U.S. and Europe. This slight increase in total energy consumption is largely due to modest population growth. By comparison, total energy consumption in Asia has increased considerably between 1980 and the present. The increase in energy use from this region is the principal reason for the increase in overall world energy consumption. The increase in energy consumption in Asia (in particular) is due mainly to significant increases in energy use in India and China. This increase in energy use is due to population increases and, most significantly, the increase per capita in energy consumption due to an increased level of industrialization. This observation is important in consideration of future energy needs as discussed in what follows.

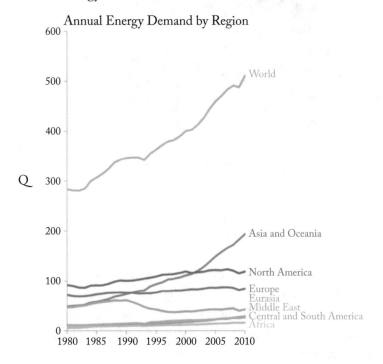

Figure 1.2: Total annual energy use for different regions from 1980–2010 (Q = quad = 1015 Btu = 1.055×10^{18} J). https://commons.wikimedia.org/wiki/File:World_primary_energy_consumption_in_quadrillion_Btu_by_region.svg. **Based on Martin Kraus/wikimedia commons/CC BY-SA.3.0.** https://creativecommons.org/licenses/by-sa/3.0/deed.en.

1.3   OUR ENERGY SOURCES

The breakdown of primary energy sources in the U.S. for 2017 is shown in Figure 1.3. It is clear that the vast majority of energy (over 80%) comes from various forms of fossil fuels. Historically this has been the case, as shown in Figure 1.4, where it is seen that fossil fuels have continued to dominate U.S. energy production and only modest increases in nuclear energy and renewables has occurred since around 1970.

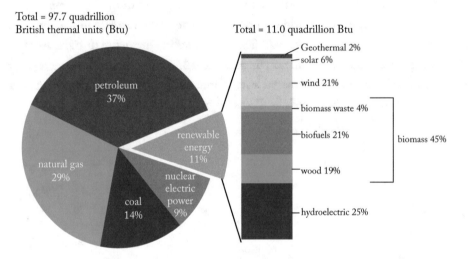

Figure 1.3: Breakdown of primary energy use by source in the U.S. in 2017 (1 quadrillion Btu = 1.055 × 10^{18} J). Based on https://www.eia.gov/energyexplained/?page=us_energy_home.

Worldwide energy use is fairly similar to that in the U.S. and is summarized in Table 1.1. The total fossil fuel use is very similar to the U.S., although the proportions of petroleum, natural gas, and coal are somewhat different. Worldwide, the use of coal is more prevalent than in the U.S. and this is largely the result of extensive coal use in Asia (and particularly in China) and the extensive overall energy use in this region of the world, as seen in Figure 1.2.

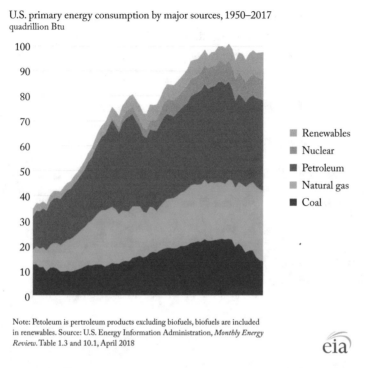

U.S. primary energy consumption by major sources, 1950–2017
quadrillion Btu

Note: Petoleum is pertroleum products excluding biofuels, biofuels are included
in renewables. Source: U.S. Energy Information Administration, *Monthly Energy
Review.* Table 1.3 and 10.1, April 2018

Figure 1.4: Breakdown of U.S. primary energy use by source since 1950 (1 quadrillion Btu = 1.055 × 1018 J). Based on https://www.eia.gov/energyexplained/?page=us_energy_home.

Table 1.1: World primary energy sources for 2016. Coal includes peat and oil shale; others includes geothermal, solar, wind, tidal, and other renewables. Data are from https://www.iea.org/statistics/kwes/supply/

Source	Percent Total Energy
Petroleum	31.9
Coal	27.1
Natural gas	22.1
Biofuels	9.8
Nuclear	4.9
Hydroelectric	2.5
Other	1.6

A breakdown of end-user energy compared with primary energy use is illustrated in Figure 1.5. An important fact illustrated in this chart is the fraction of primary energy that contributes to the production of electricity. The chart shows that about 38% of primary energy is used for electricity generation and that only about 34% of the primary energy that goes into electricity production actually produces usable electricity; the remaining 66% is rejected energy in the form of thermal energy given off to the environment. It is also interesting to observe from the chart that virtually all renewable energy (except for biomass energy) goes into electricity production.

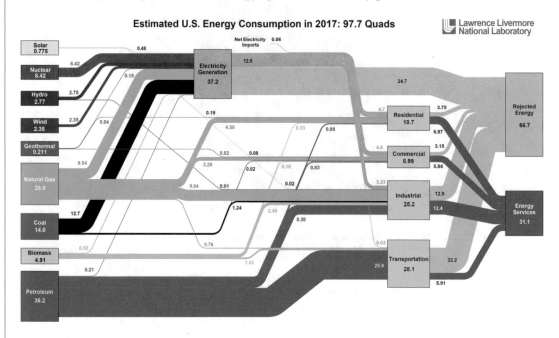

Figure 1.5: Flow chart of U.S. energy use in 2017 showing the relationship of end-used energy to primary energy source. (1 Quad = 1 quadrillion Btu = 1.055 × 1018 J). Source : LLNL and DoE, https://flowcharts.llnl.gov/commodities/energy.

While there is a great deal of similarity between energy production methods in industrialized countries, there are also some significant differences. The generation of electricity is a significant example of how national resources and energy policies contribute to a country's energy production. Table 1.2 gives a breakdown of electricity-generating methods used by some European nations. As can be seen from the data, the fraction of electricity generated using fossil fuels ranges from virtually zero to well over 90%, while the fraction of electricity produced by renewable sources ranges from a few percent to nearly 100%. The importance of nuclear power is also seen in the table as ranging from zero to approximately three quarters of electricity generation. Many of these variations are the result of resource availability, such as in Iceland, which has extensive geothermal and hydroelectric resources and Norway, which has extensive hydroelectric resources, as well as Estonia, which has considerable oil shale deposits. However, other features that are shown in the table are the result of national energy policy, such as France's choice to promote the use of nuclear power.

Another interesting example is a comparison of electricity generation in Canada and the U.S., as summarized in Table 1.3. Both countries have similar land area, considerable fossil fuel resources and similar hydroelectric and wind resources. The U.S. has greater solar energy resources because of its latitude. Both countries make use of nuclear energy. It is seen, however, that there are major differences between the two countries in their use of fossil fuels and renewable energy for electricity generation. While the U.S. depends primarily on fossil fuels for its electricity, Canada obtains most of its electricity from renewables, particularly hydroelectricity. The total installed hydroelectric capacity and the total annual hydroelectric generation is similar between these two countries, as seen in Table 1.4. However, the greater population in the U.S., and the subsequent greater total electricity requirement, means that hydroelectric energy accounts for a smaller percentage of the total electricity use and, in this case, additional electrical requirements are met largely by fossil fuels. This is an example of the effects of population density of the utilization of renewable energy resources.

Table 1.2: Breakdown of electricity generation in European countries for August 2015 to August 2016. Countries are organized by increasing contribution of fossil fuels to electricity production (data are from https://icelandmag.is/article/iceland-meets-only-001-its-electricity-needs-fossil-fuels-9999-renewables)

Country	Fossil Fuels (%)	Nuclear (%)	Renewables (%)
Iceland	0.01	0.00	99.99
Norway	2.13	0.00	97.87
France	7.79	74.48	17.73
Sweden	8.87	37.41	53.73
Luxembourg	21.50	0.00	78.50
Slovakia	23.48	57.68	18.84
Austria	27.81	0.00	72.19
Slovenia	29.44	38.76	31.80
Finland	36.48	34.22	29.29
Belgium	37.42	49.45	13.13
Spain	39.85	21.05	39.10
Hungary	42.96	53.09	3.94
Portugal	47.05	0.00	52.95
Denmark	49.95	0.00	50.05
Germany	62.09	13.10	24.81
Czech Republic	62.41	31.47	6.12
United Kingdom	63.35	19.47	17.19
Italy	66.46	0.00	33.54

Greece	70.00	0.00	30.00
Ireland	74.35	0.00	25.65
Netherlands	87.51	3.50	8.99
Poland	90.53	0.00	9.47
Estonia	93.61	0.00	6.39

Table 1.3: Breakdown (in %) of primary sources of energy for electricity generation in Canada and the U.S. (data for Canada from https://www.nrcan.gc.ca/energy/facts/electricity/20068; data for U.S. from https://www.eia.gov/tools/faqs/faq.php?id=427&t=3)

Energy Source	Canada	United States
Natural gas	9.2	31.7
Coal	9.0	30.1
Petroleum	1.0	0.5
Other fossil fuels	~0	0.4
Fossil fuels total	19.2	62.7
Nuclear	14.7	20.0
Hydroelectric	59.1	7.5
Wind	4.7	6.3
Biomass	1.7	1.6
Solar	0.5	1.3
Geothermal	~0	0.4
Renewables total	66.0	17.1
Other	~0.1	~0.2

Table 1.4: Comparison of hydroelectric capacity and generation in Canada (for 2016) and the U.S. (for 2017) (data are from https://www.hydropower.org/country-profiles; https://www.nrcan.gc.ca/energy/facts/electricity/20068; https://www.statista.com/statistics/188521/total-us-electricity-net-generation/)

Country	Population (10^6)	Land Area (km^2)	Population Density (km^{-1})	Hydroelectric Installed Capacity (GW)	Hydroelectric Energy Generated (TWh)	Capacity Factor (%)	Total Electric Generation (TWh)	Percent Hydroelectric (%)
Canada	36.3	9.98×10^6	3.6	79	380	55	648	59
United States	325.7	9.83×10^6	33.1	103	322	36	4015	8

Transportation energy poses unique problems as, in most cases, the source of energy to propel a vehicle must be portable. A breakdown of transportation energy use in the U.S. is shown in Figure 1.6. In the U.S., transportation accounts for 29% of all primary energy use. While rail transportation may, in some cases, utilize an external source of energy, the other modes of transportation must generally carry the energy source with them. This source of energy must, in the case of road transportation and (particularly) air transportation, be of a sufficiently high energy density as to allow for the construction of a vehicle within reasonable mass limits. As shown in Figure 1.6, liquid or gaseous fossil fuels are, by far, the major source of energy for transportation, at present. A similar situation exists for other industrialized countries, as illustrated for Canada in Figure 1.7.

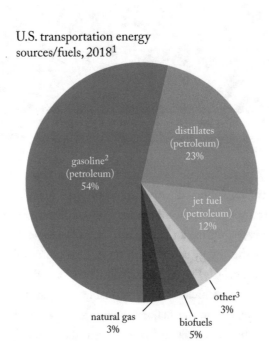

U.S. transportation energy
sources/fuels, 2018[1]

distillates
(petroleum)
23%

gasoline[2]
(petroleum)
54%

jet fuel
(petroleum)
12%

other[3]
3%

natural gas
3%

biofuels
5%

[1] Based on energy content

[2] Motor gasoline and aviation gas; excludes ethanol

[3] Includes residual fuel oil, lubricants, hydrocarbon gas liquids (mostly propane),
and electricity (includes electrical system energy losses).

Note: Sum of individual components may not equal 100% because of
independent rounding.

Source: U.S. Energy Information Administration, *Monthly
Energy Review*, Tables 2.5, 3.8c, and 10.2b, April 2019,
preliminary data

eia

Figure 1.6: Breakdown of sources of energy for transportation in the U.S. in 2017. Note: "distillates" include (primarily) diesel fuel; "Others" includes (primarily) propane and electricity. Based on https://www.eia.gov/energyexplained/?page=us_energy_transportation.

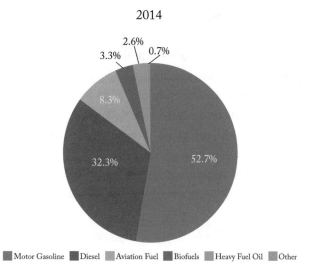

2014

Figure 1.7: Transportation fuel breakdown for Canada for 2014. (Note: "Heavy fuel oil" is used in marine and rail transportation. Biofuels include ethanol and biodiesel blended with petroleum products. "Other" includes natural gas, electricity, lubricants and propane). Based on and reproduced with the permission of Public Works and Government Services, (2019). https://www.neb-one.gc.ca/nrg/ntgrtd/ftr/2016/index-eng.html.

1.4   OUR FUTURE ENERGY NEEDS

The total world energy need in the future depends on several factors. In simple terms, the total world energy use can be expressed as the product of the world population and the average per capita energy use. The world population has increased consistently over time and presently sits at around 7 billion. This number will surely increase in the future, but it cannot continue to increase indefinitely. Populations of species in the wild are self-limiting and depend on the availability of suitable habitat and food. The same is true of humans. The population will be limited by the ability to produce food. Many studies suggest that a population around 10 billion would be the maximum sustainable.

The average per capita power consumption worldwide is about 2.7 kW, compared to around 12 kW in industrialized countries. A simple approach to estimating the maximum possible energy needs of society in the future would be to assume that the population reached its sustainable maximum and that the per capita energy consumption would be the same worldwide as it currently is in industrialized countries. This approach would give a limiting maximum energy consumption of about 5.8 times the current value of 6×10^{20} J/y or about 3.5×10^{21} J/y. While it is possible that this value will be even greater as a result of future energy needs that we do not currently envision, it is more likely that conservation efforts (perhaps due, at least in part to resource limitations) will actually reduce this value. There is also no certainty of worldwide energy equality a century from now. In fact, some estimates predict that total world energy use early in the 22nd century will be more-or-less the same as it is at present. The most probable estimates fall somewhere in the middle of this range, probably around 1.5×10^{12} J/y. Even predictions on a substantially shorter time scale are subject to considerable variability. For example, a study by the United Nations predicts a 50% increase in total world energy use between 2014 and 2035, while the U.S. Energy Information Administration predicts a 28% increase between 2017 and 2040. In any case, it is important to keep in mind that future energy needs will likely be greater than they are at present.

While assessing total future energy needs is difficult, predicting actual sources of energy in the future is even more challenging and predictions beyond 20–30 years are not likely to be very meaningful.

Figure 1.8 shows one prediction of U.S. primary energy sources until 2040. This prediction shows that petroleum, nuclear, hydroelectric, and biofuels will remain more-or-less constant in the near future, while renewables and natural gas will increase and coal will decrease. Although the overall energy production from fossil fuels remains more-or-less constant, greenhouse gas emissions will decrease by about 4% between 2016 and 2044, because coal produces more CO_2 per unit energy than natural gas. Figure 1.9 shows that worldwide coal use is expected to remain more-or-less constant (because of its extensive coal use in Asia), while energy from all other sources will increase.

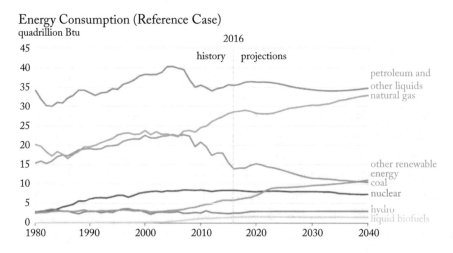

Figure 1.8: U.S. total primary energy use since 1980 and predictions until 2040. Based on https://www.eia.gov/outlooks/aeo/pdf/0383(2017).pdf, the U.S. Energy Information Administration (January 2017).

IEO2018 Reference Case
World energy consumption by energy source

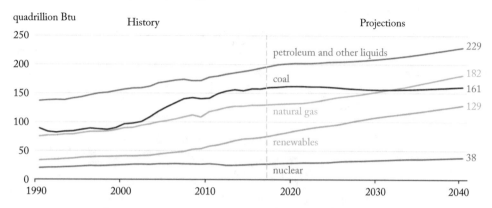

Figure 1.9: World total primary energy use since 1990 and predictions until 2040. Based on https://www.eia.gov/pressroom/presentations/capuano_07242018.pdf, the U.S. Energy Information Administration (July 2018).

Predictions for electricity generation in the U.S. are shown in Figure 1.10. Trends are similar to those shown for overall U.S. primary energy consumption, as shown in Figure 1.8. As noted above, about 38% of the primary energy used in the U.S. is converted into electricity and this percentage is similar in most industrialized countries. Figure 1.11 shows the projected sources of electricity generation in China until 2040. The figure shows that coal use for electricity generation in China will remain relatively constant over the next two decades or so. The additional electricity requirements will be made up by increased nuclear and renewable energy.

As indicated in Figure 1.6, current energy sources for transportation are weighted heavily toward fossil fuels. One might expect, therefore, if renewable energy sources are actively developed in the foreseeable future, that this trend would most noticeably be seen in transportation fuels. This, however, is not necessarily the case. Figure 1.12 shows projected transportation energy sources for the U.S. until 2040. During this period there is relatively little change in fossil fuel use and, perhaps surprisingly, electricity remains a fairly insignificant component of transportation energy even twenty years from now. This feature is also illustrated by the prediction shown in Figure 1.13 for new light duty vehicle sales in the U.S. By 2040 battery electric and fuel cell vehicles (i.e., those that do not require fossil fuels) are expected to account for about 1 million vehicle sales annually. This may be compared to total new light duty vehicle sales of about 17 million in 2017. Thus, substantially less than 10% of new vehicles are expected to be fossil fuel free.

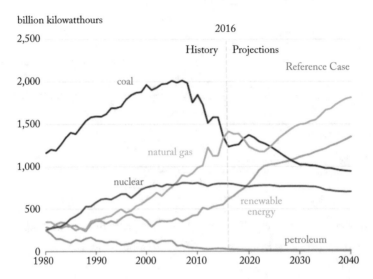

Figure 1.10: U.S. primary energy used for electricity generation since 1980 and predictions until 2040. Based on   https://www.eia.gov/outlooks/aeo/pdf/0383(2017).pdf, the U.S. Energy Information Agency (January 2017).

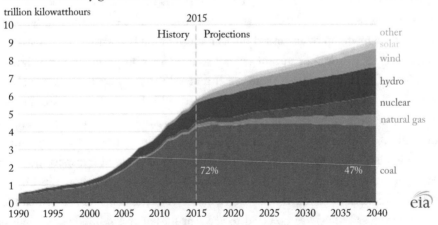

Figure 1.11: Projected sources for electricity generation in China. Based on https://www.eia.gov/today-inenergy/detail.php?id=33092. Source: U.S. Energy Information Administration (September 2017).

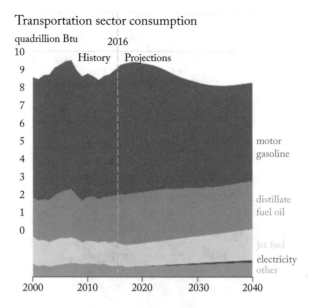

Figure 1.12: U.S. primary energy used for transportation since 1980 and predictions until 2040. Based on https://www.eia.gov/outlooks/aeo/pdf/0383(2017).pdf, the U.S. Energy Information Agency (January 2017).

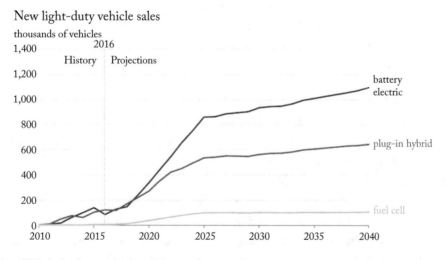

Figure 1.13: U.S. light duty vehicle sales since 2010 and projections to 2040 for battery electric, plug-in hybrid and fuel cell vehicles. Based on https://www.eia.gov/outlooks/aeo/pdf/0383(2017).pdf, the U.S. Energy Information Agency (January 2017).

A similar situation is seen for predicted transportation fuels in Canada, as seen in Figure 1.14. A comparison with the present situation illustrated in Figure 1.7 shows only a minimal reduction in fossil fuel consumption for transportation use in Canada over the next 20 years or so.

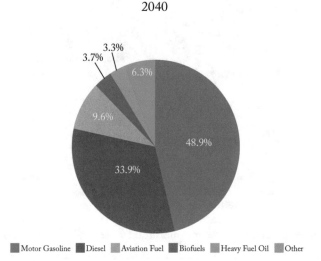

Figure 1.14: Projected transportation fuel breakdown for Canada for 2040. Estimates from the National Energy Board of Canada. (Note: Hheavy fuel oil is used in marine and rail transportation. Biofuels include ethanol and biodiesel blended with petroleum products. Other includes natural gas, electricity, lubricants and propane). Source: Canada's Energy Future 2016: Energy Supply and Demand Projections to 2040 (January 2016). Based on and reproduced with the permission of Public Works and Government Services, (2019) https://www.neb-one.gc.ca/nrg/ntgrtd/ftr/2016/index-eng.html.

While the situation illustrated in Figures 1.12 and 1.14 is, on average, representative of expected transportation fuel sources worldwide, there are countries where there are more proactive approaches to fossil fuel reduction. Iceland, for example, is taking a very aggressive approach to eliminating carbon emissions. As seen in Table 1.2, this country produces virtually all of its electricity from renewable sources (~73% hydroelectric and ~27% geothermal). According to a recently adopted energy policy, the registration of new fossil fuel burning vehicles will be banned in Iceland as of 2030 and the use of fossil fuels for transportation will be eliminated by 2050.

The above discussion provides some insight into our current energy use and expectations for our future energy needs. It is certainly suggested that any transition away from the predominant use of fossil fuels would occur very slowly on a time scale of the next 20–30 years. It is important to consider these predictions in the context of the arguments presented in the next chapter for changing our approach to energy.

CHAPTER 2

The Need for Renewable Energy

In this chapter the reasons for changing our energy sources are considered. These reasons fall into two broad categories: diminishing resources and environmental impact. The chapter considers the Hubbert model as a means of projecting availability of fossil fuel energy resources in the future. The environmental impact of fossil fuel use is described, with particular emphasis on global climate change.

2.1 INTRODUCTION

As shown in Chapter 1, our sources of energy have changed over the years. In pre-industrial times, biofuels (specifically wood) were the major source of energy and these were used for heat and cooking. During the industrial revolution, coal became the predominant source of energy and was needed to fulfill the needs of industry and rail transportation. In the mid-20th century, oil surpassed coal, as its liquid form provided a convenient fuel for road transportation. Since that time, fossil fuels in the form of oil, natural gas, and coal comprised the vast majority of primary energy use worldwide. However, during latter part of the 20th century it became increasingly obvious that the continued use of fossil fuels as our major energy source was neither sustainable, in terms of the availability of resources, nor environmentally acceptable. The present chapter overviews some of the approaches to estimating the longevity of remaining fossil fuel reserves, as well as the evidence for the environmental consequences for their continued use.

2.2 RESOURCE LIMITATIONS

2.2.1 RESERVES-TO-PRODUCTION RATIOS

Perhaps the simplest approach to estimating the longevity of existing fossil fuel resources is based on a consideration of the remaining world reserves and their current rate of use. This is the so-called "reserves-to-production ratio" and this ratio is a direct measure of the remaining lifetime of the resource. This approach makes the assumption that current estimates of remaining viable resources are correct and that additional extensive reserves are not discovered. It also makes the assumption that use will be constant at the current level until the resource is completely depleted. The first assumption is based on our understanding of the availability of fossil fuel deposits world-wide and their economic viability. A careful consideration of geological resources should provide a reasonable assessment of total reserves. This is considered further in the following discussion of the Hubbert model. The second assumption is clearly not an accurate description of resource utilization. There may be increased fossil fuel consumption to accommodate increasing energy needs, at least to the level that can be accommodated by the existing infrastructure. On the other hand, at some point there is likely to be decreased fossil fuel consumption due to diminishing resources or shifts in energy policy to other, more environmentally conscious, alternatives. However, this simple approach can provide a first estimate of fossil fuel lifetimes that can serve as a basis for comparison to other models.

Table 2.1 gives results of the simple reserves-to-production ratio analysis for world oil, natural gas, and coal. Oil and natural gas show a lifetime of about 50 years (until around 2070) and coal shows a lifetime of about 100 years (until around 2120).

Table 2.1: Reserves-to-production ratios (R/P) for oil, natural gas, and coal. Reserves data estimates are from BP statistical review (2015): t = metric tonnes. (Note: Oil is sometimes measured in tonnes rather than liters or barrels because different oils can have different densities.)

Fuel	Reserves	Current Annual Production	R/P (years)	End Year
Oil	2.18×10^{11} t	4.46×10^9 t	49	2064
Natural gas	1.90×10^{14} m^3	3.54×10^{12} m^3	54	2069
Coal	8.07×10^{11} t	7.70×10^9 t	105	2120

2.2.2 HUBBERT THEORY

A more detailed mathematical model for the determination of the longevity of fossil fuel resources was proposed in the mid-1950s by the American geophysicist Marion King Hubbert. This model has been used extensively for the analysis of energy resources, but is also applicable to other natural resources, as well. The basic assumptions made by Hubbert in the development of this model are as follows.

1. The production rate of a natural resource begins slowly as new sources of the resource are discovered and infrastructure for the extraction and utilization of the resource are established.

2. The production rate will reach a maximum when half of the resource has been utilized.

3. The production rate will go to zero as the resource is exhausted.

The model includes the requirements that the exploration effort is constant over the lifetime of the resource and that the time between discovery and production is a constant. These assumptions lead to a cumulative production curve as a function of time, $Q(t)$, as shown in Figure 2.1(a). The cumulative production curve shows a "sigmoidal" shape and approaches a value of $Q\infty$, the total quantity of the resource that is commercially viable. Although the Hubbert model does not imply any specific mathematical form of the cumulative production curve, it is commonly assumed to be expressed mathematically as the logistic growth curve:

$$Q(t) = \frac{Q_\infty}{1 + e^{a(b-t)}} .$$ (2.1)

It is easy to see that $Q(b) = Q_\infty/2$, meaning that half of the resource will be used at a time $t = b$. It is often convenient to plot the production rate, $P(t)$ (rather than the cumulative production), as a function of time. The production rate is given as the derivative of the cumulative production as

$$P(t) = \frac{dQ_\infty(t)}{dt} .$$ (2.2)

Differentiating Equation (2.1) leads to

$$P(t) = \frac{aQ_\infty}{\left(e^{-a(b-t)/2} + e^{a(b-t)/2}\right)^2} .$$ (2.3)

A plot of this function, known as the logistic probability distribution, is shown in Figure 2.1(b) and illustrates the assumptions of the Hubbert model. Although this curve resembles a Gaussian, it decreases more slowly in the tail sections. The maximum production rate, P_{max}, occurs for $t = b$, the time when half of the resource has been used, and Equation (2.3) gives

$$P_{max} = \frac{aQ_\infty}{4} .$$ (2.4)

The parameter a is inversely related to the width (in time) of the production rate curve in Figure 2.1(b). The analysis of production data for specific energy resources on the basis of the above theory allows for an understanding of the time dependence of the production rate and an evaluation of the longevity of the resource.

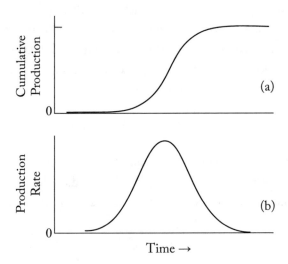

Figure 2.1: (a) The logistic growth curve used to describe cumulative production in the Hubbert model and (b) logistic probability distribution used to describe the time dependence of the production rate.

The application (and validity) of the Hubbert model can be seen by applying the technique to a limited set of localized data. One example of this approach is illustrated in Figure 2.2, where the time dependence of the production of Pennsylvania anthracite is shown. Anthracite is a variety of coal that has a particularly high carbon content and minimum impurities and, therefore, has a higher energy content and produces fewer pollutants than other varieties of coal. In the U.S. there are (or were) notable anthracite deposits in Pennsylvania and these were extensively mined in the early period of coal utilization. These deposits were virtually exhausted by around 1980 and anthracite is an example of a specific resource that complies with the assumptions of the Hubbert model. The graph of anthracite production shows the characteristic bell-shaped curve that results from the Hubbert model. A slight anomaly is seen around the period of World War II, when there was increased production to meet energy demands.

The application of the Hubbert model to world fossil fuel production will provide an assessment of the long-term availability of these resources and future expectations for production. Figure 2.3 shows a Hubbert analysis for world oil, natural gas, and coal production data up to the present and an extrapolation of the curves to describe future production. It is seen that the historical data all show a shape that corresponds well to the shape of the logistic curve with the exception of a small anomaly along the lines of the anomaly in the Pennsylvania anthracite data. This anomaly is most obvious in the increased production in oil that occurred during the 1970s. The data presented in Figure 2.3 have, therefore, been fitted to a multi-Hubbert function, to account for increased production at some point in the past.

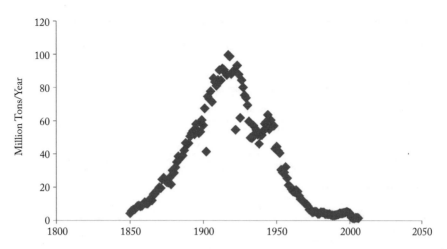

Figure 2.2: Production rate for Pennsylvania anthracite from the mid-1800s to around 2000. Based on Brecha, 2012.

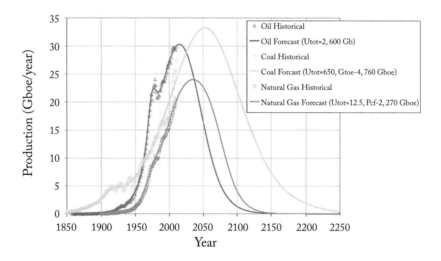

Figure 2.3: Multi-Hubbert analysis of world oil, natural gas, and coal. Based on Maggio and Cacciola (2012), reprinted with permission from Elsevier. https://doi.org/10.1016/j.fuel.2012.03.02.

The analysis of the data in Figure 2.3 provides values for the peak production year for the three different fossil fuels as given in Table 2.2. Based on the width of the distribution, the time at which production will drop to one half of the peak value can also be determined, as given in the table. For a single logistic curve, this time corresponds to the consumption of 85.3% of the total resource. Values of the half peak production year from Table 2.2 show trends that are consistent with the end years for the simple reserves-to-production ratios, as given in Table 2.1. Certainly, by the time 85% of fossil fuel resources are consumed, it will be necessary to have the infrastructure for alternative energy sources available to supplement the reduced fossil fuel production.

Table 2.2: Years corresponding to peak world production of oil, natural gas, and coal from the multi-Hubbert analysis; data from Maggio and Cacciola (2012). The half-peak production year is obtained from the data in Figure 2.4 and corresponds to the time when future production will drop to one half of the peak production

Fuel	Peak Year	Half-peak Year
Oil	2014	2053
Natural gas	2032	2077
Coal	2051	2111

An interesting feature of Hubbert analysis can be seen by looking at oil production in the U.S. This is shown in Figure 2.4. From the beginning of oil recovery in the U.S. in the late 1800s, until around 2010, the data show the general shape of the Hubbert logistic curve and, in the past, these data have been used as a means of validating the Hubbert model. In fact, when Hubbert developed his theory in the mid-1950s, he specifically sought to predict future U.S. oil production.

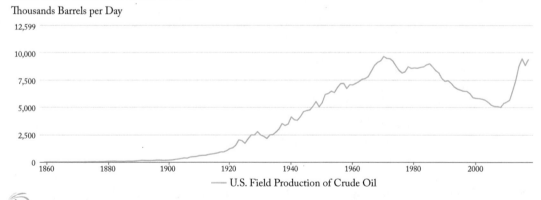

Figure 2.4: Production rate of crude oil in the U.S. from the late 1800s to the present. Based on https://www.eia.gov/dnav/pet/hist/LeafHandler.ashx?n=PET&s=MCRFPUS2&f=A, the U.S. Energy Information Administration.

Prior to around 2010, oil produced in the U.S. was largely what is now referred to as "conventional" oil, that is, oil produced from conventional oil wells by conventional methods. If only conventional oil production is plotted since around 2010, the curve will continue to follow the logistic curve predicted by Hubbert theory. The increased production since about 2010 is from non-conventional oil. This oil is sometimes referred to as "tight oil." It refers to oil that is more difficult to extract and may include oil obtained by non-conventional techniques, such as fracking, or oil that comes from oil shale. These techniques are typically less environmentally acceptable than the conventional techniques used for extracting fossil fuel resources. A similar up-turn in production of U.S. natural gas is observed on a similar time scale as a result of the implementation of fracking technology in that industry.

As a result of the implementation of non-conventional technology for the extraction of oil in the U.S., an increase in the total available resource in the Hubbert model is necessary. The resulting bimodal production curve follows along the lines of the multi-Hubbert analysis previously shown which included increased production during times of increased demand.

While the effect of non-conventional oil on U.S. production was quite notable during the past decade, it is important to consider its significance on a global scale. Figure 2.5 shows the results of a model that incorporates estimated global non-conventional oil resources. The peak year and half-peak year are consistent with those of the model for world oil shown in Figure 2.3. However, the tail resulting from non-conventional oil extends somewhat further into the future.

THE HUBBERT CURVE

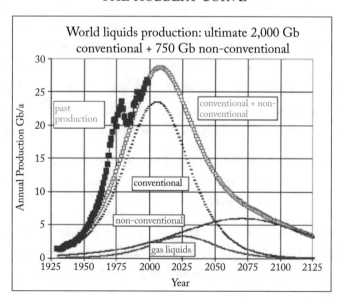

Figure 2.5: Hubbert analysis of world oil showing estimates for conventional and non-conventional resources. Laherrère (2001), available at http://www.hubbertpeak.com/laherrere/iiasa_reserves_long. pdf. Based on and reproduced with permission of J. Laherrere.

2.3   ENVIRONMENTAL CONSEQUENCES OF OUR ENERGY USE

The utilization of energy by humanity inevitably has adverse environmental effects. These effects may have short-term, local implications or they may have global consequences that persist over a long (or indefinite) period of time. As fossil fuels currently account for more than 80% of our energy use, this section considers the environmental effects of our use of oil, natural gas, and coal.

2.3.1   POLLUTION

Pollution that results from the combustion of fossil fuels may be in the form of thermal pollution or chemical pollution as described below.

Heat produced by the combustion of fossil fuels may be converted into mechanical energy (and subsequently, in some cases, to electricity) by means of a heat engine. As the thermodynamic efficiency of a heat engine is always less than 100%, excess heat is released to the environment. This situation is most notable for vehicles utilizing internal combustion engines, where the thermodynamic efficiency is typically less than about 20% and thermal generating stations to produce electricity where the thermodynamic efficiency of typically less than 40%. Thermal pollution, therefore, results in regions where there is a concentration of vehicles and around electric generating stations. In the latter case, the transfer of excess thermal energy to the environment is either through the use of cooling water from rivers, lakes, or the ocean or to the atmosphere through cooling towers. Thermal pollution of waterways can have adverse effects on aquatic wildlife, while the use of cooling towers can contribute to undesirable local atmospheric conditions, such as regions of excess fog.

Chemical pollution from the combustion of fossil fuels can take several forms.

- Carbon monoxide and hydrocarbons result from incomplete combustion of the fuel. These are most common in the exhaust products of internal combustion vehicles.

- Nitrogen-oxide compounds (NO_x) result from the oxidation of nitrogen in the air and these occur in any combustion process. Again, these are most notable in the exhaust from internal combustion vehicles.

- Sulfur dioxide results from the presence of sulfur impurities in fossil fuels. This is most notable for the combustion of coal in electric generating stations. Sulfur dioxide reacts with components of the atmosphere to produce sulphuric acid, which results in so-called acid rain.

- Particulate matter pollution results from the combustion of solid fuels, most notably coal.

- Carbon dioxide results from the combustion of any carbon containing fuel.

The health effects of pollution caused by fossil fuel combustion are a serious concern, especially in cases where pollutants are concentrated as a result of adverse atmospheric conditions. However, these are largely local, and in many cases, have short-term effects. Much progress has been made in recent years in reducing these pollutants. Improvements in automobile engine design, as evidenced by the emission control standards shown in Table 2.3, have done much in reducing vehicle emissions. Coal-fired generating stations have substantially reduced sulfur dioxide emissions using scrubbers (where exhaust gas is reacted with CaO or $CaCO_3$) and particulate pollution using filters.

Table 2.3: Progress in eliminating pollutants from vehicle exhaust between 1960 (pre-emission control average) and 2004 U.S. emission standards (values in g/km)

Year	CO	NO_x	Hydrocarbons
1960	52.2	2.5	6.58
2004	1.1	0.04	0.056

The most serious effect of fossil fuel use is the emission of greenhouse gases (i.e., CO_2), as described below, as this has long-lasting global consequences.

2.3.2   THE GREENHOUSE EFFECT

The temperature of the surface of the earth is determined by the intensity of radiation it receives from the sun and the nature of its atmosphere. While there is also a small component of heat from geothermal sources, this is less than 0.1% of the total heat flow and is not significant. The intensity of the solar radiation incident on the outside of the earth's atmosphere, S, is given in terms of the total power output of the sun, P, and the distance between the earth and sun, R, as

$$S = \frac{P}{4\pi R^2} \ .$$

(2.5)

Using the values $P = 3.8 \times 10^{26}$ W and $R = 1.49 \times 10^{11}$ m gives a value of $S = 1.367$ kW/m². This is often referred to as the solar constant. The equilibrium surface temperature of the earth can be easily calculated from the solar constant for the case where there is no atmosphere present by equating the incident radiation absorbed to the power radiated back into space. The absorbed incident solar power on the earth (which appears as a disk of area πR_e^2, R_e = earth radius) will be

$$P_{incident} = \pi R_e^2 \, (1 - a)S \ .$$

(2.6)

The quantity a is the earth's albedo, that is the fraction of incident radiation that is reflected. The quantity $(1 - a)$ is, therefore, the fraction of radiation that is absorbed.

The earth will heat to an equilibrium temperature, T, and radiate energy as a blackbody back into space according to the Stefan-Boltzmann law

$$P_{radiated} = 4\pi R_e^2 \, \sigma T^4 .$$ (2.7)

where σ is the Stefan-Boltzmann constant; $\sigma = 5.67051 \times 10^{-8}$ W·m^{-2} · K^{-4}. The equilibrium surface temperature of the earth is obtained by equating the absorbed incident power and the radiated power. This gives the temperature

$$T = \left[\frac{(1-a)S}{4\sigma} \right]^{1/4} .$$ (2.8)

Using the solar constant as given above and the measured earth's albedo ($a = 0.3$), the equilibrium temperature is found to be 254 K (or about -19°C). This value is lower than the known average surface temperature of the earth (currently about 288 K) and is also inconsistent with the development of life. The discrepancy between the temperature calculated above and the actual temperature is due to the presence of the earth's atmosphere and the accompanying greenhouse effect as described below.

The sun may be approximated by a blackbody with a surface temperature of 5,800 K. The peak in the spectral distribution of radiation from a blackbody corresponds to a wavelength as given by Wien's displacement law:

$$\lambda_{max} = \frac{b}{T} .$$ (2.9)

where b is the Wien displacement constant; $b = 2.8978 \times 10^{-3}$ m · K^{-1}. The peak in the spectrum of the sun is, therefore, at a wavelength of 5×10^{-7} m or 500 nm, which is in the green portion of the optical part of the electromagnetic spectrum.

The radiation which is emitted by the surface of the earth corresponds to a blackbody with a temperature of around 300 K. From the Wien displacement law this radiation would have a wavelength of about 9.7×10^{-6} m or 9,700 nm. This wavelength is well into the infrared portion of the electromagnetic spectrum. The greenhouse effect results from the fact that different molecules interact differently with electromagnetic radiation as a function of wavelength. Specifically, greenhouse gases, such as CO_2, scatter more strongly with long wavelength electromagnetic radiation than with short wavelength electromagnetic radiation. This point is illustrated in Figure 2.6 which shows the radiation balance of the earth.

The incident solar radiation is seen to be 342 W/m^2 in Figure 2.6. This value is readily determined from the solar constant. From the sun the earth appears as a disk with an area of πR_e^2. The total surface area of the earth is $4\pi R_e^2$. The average solar radiation per unit area of the earth, (sometimes called the insolation), I, is, therefore,

$$I = \frac{\pi R_e^2 S}{4\pi R_e^2} = \frac{S}{4} = \frac{1367 \text{ W/m}^2}{4} = 342 \text{ W/m}^2 \,. \tag{2.10}$$

The figure shows that a portion of the incident solar radiation is reflected by either the atmosphere or the surface. This reflected radiation (77 W/m^2 + 30 W/m^2 = 107 W/m^2) is related to the earth's albedo; a = (107 W/m^2)/(342 W/m^2) ≈ 0.3. The remaining radiation, 225 W/m^2, is absorbed by either the earth's surface or atmosphere, and is responsible for warming the planet.

The earth's surface and atmosphere re-radiate energy as a blackbody at about 300 K. Greenhouse gases in the atmosphere reflect some of this long wavelength radiation back to the surface, thus reducing the amount of energy radiated into space. This effect increases the average temperature of the earth by about 35 K. Increasing the concentration of greenhouse gases in the atmosphere will increase the radiation reflected back to the surface and thereby reduce the amount that escapes into space. This further increases the equilibrium temperature of the earth's surface. Table 2.4 gives the properties of the three most important greenhouse gases in the earth's atmosphere. Experimental evidence relating greenhouse gas concentration with the earth's average surface temperature is discussed below.

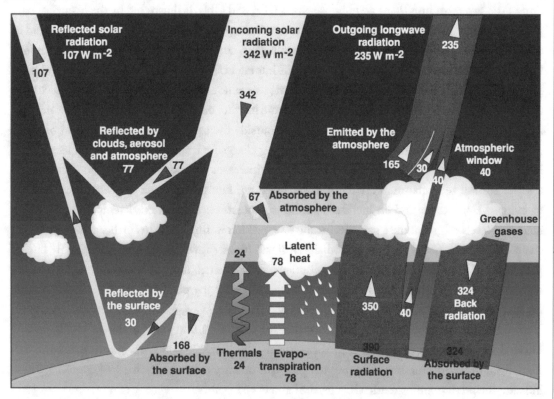

Figure 2.6: Radiation balance for the earth showing the importance of greenhouse gases in the atmosphere. Source: NASA, https://ceres.larc.nasa.gov/ceres_brochure.php?page=2.

Table 2.4: The three most important greenhouse gases in the earth's atmosphere; and CO_2 for 2016, CH_4 and N_2O for 2015. The radiative forcing is a quantitative measure of the gas's contribution to the greenhouse effect. Concentration data are from https://www.eea.europa.eu/data-and-maps/indicators/atmospheric-greenhouse-gas-concentrations-10/assessment

Chemical Species	Formula	Current Atmospheric Concentration (ppm by volume)	Radiative Forcing (W/m2)
Carbon dioxide	CO_2	402.9	1.85
Methane	CH_4	1.835	0.51
Nitrous oxide	N_2O	0.3277	0.18

2.3.3   EVIDENCE FOR GLOBAL WARMING

The relationship between the earth's average surface temperature and the atmospheric concentration of the two most important greenhouse gases, CO_2 and CH_4, is illustrated by the historical data shown in Figure 2.7. Both CO_2 and CH_4 concentrations show periodic variations with a period of the order of 100,000 years. These variations show a close correlation with temperature. The reasons for these fluctuations are clearly not the result of human activity but are the result of natural phenomena. Although the reasons for these fluctuations are not entirely clear, it has been speculated that they may correlate with variations in the cosmic ray flux on earth. Cosmic rays consist primarily of high energy charged particles and originate outside the solar system. These charged particles cause ionization in the atmosphere and the resulting charged particles attract water molecules and lead to cloud formation. The increased cloud cover increases the earth's albedo, and this leads to a lower surface temperature. Periods of low cosmic ray flux produce the opposite effect leading to reduced cloud cover and an increased surface temperature. The increased temperature leads to increased levels of greenhouse gas, specifically CO_2, which in turn leads to further temperature increase. It has been noted that the periodicity of fluctuations shown in Figure 2.7 is roughly half of the period of the sun's orbit around the galactic core. The distribution of active cosmic ray sources such as supernovae, within the galaxy can be responsible for the periodic variations in the cosmic ray flux on earth.

Data for atmospheric CO_2 concentration for the past 1,000 years is shown in Figure 2.8. These data show levels that far exceed those during the past 400,000 years shown in Figure 2.7. Data from direct measurements during the past half century or so are shown in Figure 2.9. These indicate the present atmospheric CO_2 level of about 405 ppm.

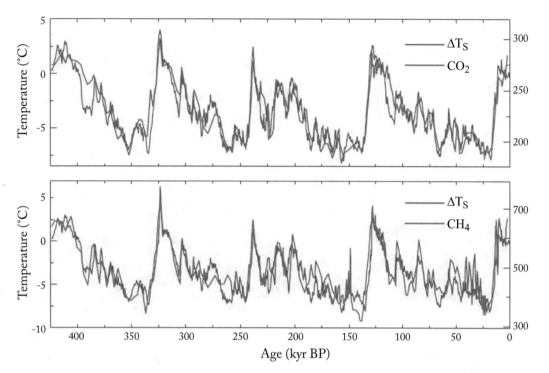

Figure 2.7: Data for atmospheric CO_2 levels (top panel) and atmospheric CH_4 levels (bottom panel) compared with temperature as obtained from Vostok, Antarctic ice core analysis. Source: NASA, https://www.giss.nasa.gov/research/briefs/hansen_11/.

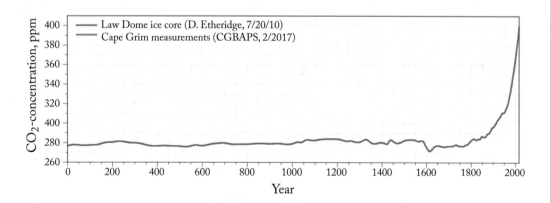

Figure 2.8: Atmospheric CO_2 levels from ice core analysis for Law Dome, Antarctic and direct measurements from Cape Grim, Australia (CC BY-SA 4.0. Based on DeWikiMan/wikimedia comons/ https://creativecommons.org/licenses/by-sa/4.0/). https://commons.wikimedia.org/wiki/File:Ghgs-lawdome-2000yr-CO2-asof2010.svg.

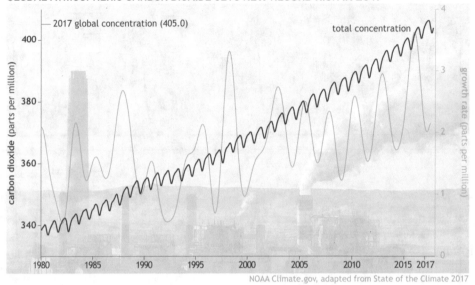

Figure 2.9: CO_2 levels at Mauna Loa Observatory in Hawaii from 1960 to present (dark red line). Annual oscillations are the result of the growth of vegetation. The light red line shows the annual growth rate in atmospheric CO_2 concentration. The graphs are overlaid on a photo of Dave Johnson Power Plant in Wyoming by Greg Goebel, used under a Creative Commons license. https://www.climate.gov/news-features/understanding-climate/climate-change-atmospheric-carbon-dioxide.

It is commonly believed that the recent increases in atmospheric CO_2 levels are the result of CO_2 emissions from anthropogenic processes, specifically the combustion of fossil fuels. CO_2 is produced by burning fossil fuels in the following way.

For the combustion of carbon, the major component of coal, CO_2 results from the reaction

$$C + O_2 \rightarrow CO_2 \; . \tag{2.11}$$

For methane, the principal component of natural gas, the reaction is

$$CH_4 + 2O_2 \rightarrow CO_2 + 2H_2O \; . \tag{2.12}$$

Oil is a mixture of complex hydrocarbons. A major component of gasoline, one of the most common refined products of oil, is octane, C_8H_{18}. The combustion of octane produces CO_2 by the reaction

$$2C_8H_{18} + 25O_2 \rightarrow 16CO_2 + 18H_2O \; . \tag{2.11}$$

The methane concentration in the atmosphere has also increased well above the historical levels of 300–700 ppb shown in Figure 2.7. This increase is illustrated in Figure 2.10. While atmospheric methane concentrations are not the direct result of the burning of fossil fuels, it is generally believed that the increase seen since the mid-19th century is the result of human activities such as increased agriculture, deforestation, waste production, and natural gas drilling. Measurements of atmospheric N_2O also show a similar increasing trend.

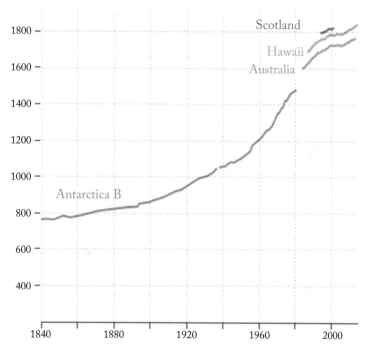

Figure 2.10: Atmospheric methane concentration in ppb since the mid-19th century. Based on https://earthobservatory.nasa.gov/Features/MethaneMatters, NASA Earth Observatory image by Joshua Stevens, using data from the EPA.

The relationship between recent atmospheric greenhouse gas concentrations and global temperature is illustrated by temperature measurements since the late 19th century as shown in Figure 2.11. The graph shows excellent agreement between the four independent data sets in the figure. A temperature increase of about 1.0°C between the early 20th century and the early 21st century is seen.

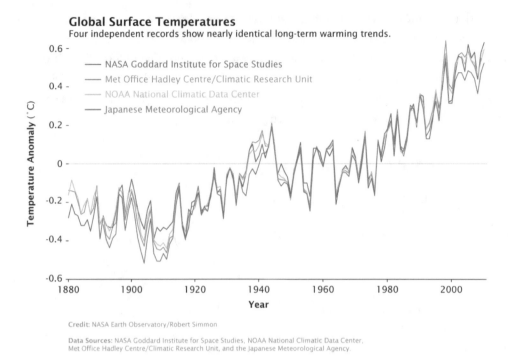

Credit: NASA Earth Observatory/Robert Simmon

Data Sources: NASA Goddard Institute for Space Studies, NOAA National Climatic Data Center,
Met Office Hadley Centre/Climatic Research Unit, and the Japanese Meteorological Agency.

Figure 2.11: Global temperature anomaly from 1880 to the present. The temperature anomaly is measured relative to the mean temperature for the years 1951–1980. Four different sets of data from different research facilities are shown (https://www.giss.nasa.gov/research/news/20110113/509983main_adjusted_annual_temperature_anomalies_final.gif).

2.3.4   CONSEQUENCES OF GLOBAL WARMING

Changes in atmospheric greenhouse gases and global temperature have a number of serious consequences. These include an increase in extreme weather events, including floods, droughts, and severe storms, and increased mortality in the elderly due to temperature extremes. Some of the direct consequences of global warming that have been clearly quantified are given below.

Increased ocean heat content. Increasing global temperatures have caused an increase in the heat being absorbed by the upper layer of the ocean. This increase in ocean heat content is clearly illustrated in Figure 2.12. These changes have significant effects on the distribution of life in the oceans.

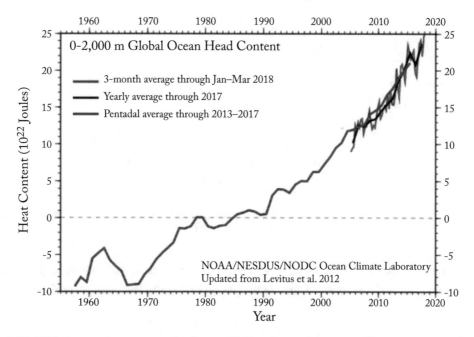

Figure 2.12: Global ocean heat content in the top 2000 m layer of the ocean. Data are measured relative to the average for the period from 1955–2006. Based on https://www.nodc.noaa.gov/OC5/3M_HEAT_CONTENT/.

Ice sheet loss. Increasing average temperatures in the Arctic and Antarctic have caused the melting of ice masses. Figure 2.13 shows measurements of the Greenland ice mass since 2002 and Figure 2.14 shows corresponding measurements for the Antarctic ice mass. In recent years, Greenland has lost an average of 286 Gt of ice per year and Antarctica has lost and average of 127 Gt of ice per year.

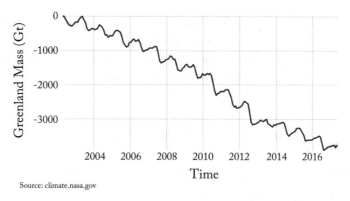

Figure 2.13: Greenland ice mass referenced to 2002 mass. Based on https://climate.nasa.gov/vital-signs/ice-sheets/.

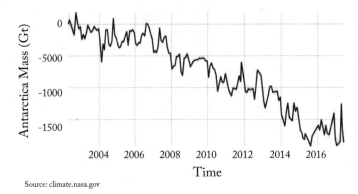

Source: climate.nasa.gov

Figure 2.14: Antarctic ice mass referenced to 2002 mass. Based on https://climate.nasa.gov/vital-signs/ice-sheets/.

Arctic sea ice decline. Rising ocean temperatures in the Arctic have resulted in the melting of sea ice and a reduction in its extent. Data for the period of 1955 to the present are shown in Figure 2.15.

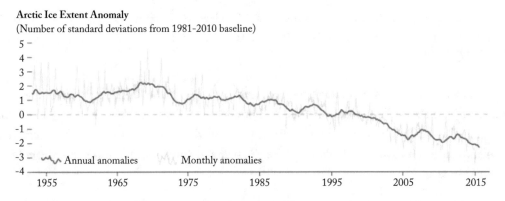

Figure 2.15: Arctic sea ice anomaly. The graph shows the quantity of Arctic sea ice above or below the average for the period of 1980–2010 in units of the standard deviation of the data from this period. Data are shown for the monthly and annual anomaly. Based on the NASA Earth Observatory graph by Joshua Stevens, using NSIDC data provided by Walt Meier/NASA Goddard, https://earthobservatory.nasa.gov/Features/SeaIce/page2.php.

Increased sea level. Melting of ice sheets and sea ice have contributed to increasing sea levels as shown in Figure 2.16. An increase of about 20 cm has been observed since the late 19th century. Such changes can have profound effects on coastal regions and continued global warming will give rise to further increases in sea level.

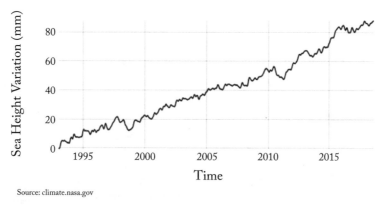

Source: climate.nasa.gov

Figure 2.16: Sea level change since 1993. Source: NASA Goddard Space Flight Center, https://climate. nasa.gov/vital-signs/sea-level/.

2.4   GLOBAL WARMING MITIGATION STRATEGIES

Data presented above show the relationship between greenhouse gas concentrations in the atmosphere and global temperature. It is the consensus of the vast majority of climate scientists that recent changes in greenhouse gas concentrations, particularly CO_2, and global warming are largely the result of anthropogenic emissions. The continued use of fossil fuels will lead to further increases in atmospheric greenhouse gas concentrations and global temperatures. Other human activities also contribute to greenhouse gas levels as is seen from rising methane levels. Activities such as deforestation reduce natural carbon sequestration and are a contributing factor, as well. While sequestering carbon from fossil fuel burning has been promoted as a possible means of mitigating greenhouse gas emissions, this a formidable task and the longevity of the sequestration is, in many cases, not certain.

Numerous studies have been undertaken to understand the future levels of greenhouse gases and their resulting effect on global temperature for different future approaches to using fossil fuels. Typical results are illustrated in Figure 2.17. The graph shows that continued use of fossil fuels according to existing policies will give rise to substantially increased greenhouse gas concentrations in the atmosphere on a time scale of a few decades. Even for the scenario where greenhouse gas emissions are reduced, the atmospheric greenhouse gas concentration will continue to rise. If greenhouse gas emissions are stopped entirely it may require more than a century for atmospheric CO_2 levels to return to pre-industrial levels.

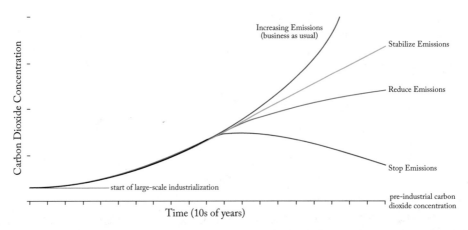

Figure 2.17: Projected atmospheric CO_2 concentration trends for different emissions scenarios. Note the scale markers on the horizontal axis are 10 years. https://earthobservatory.nasa.gov/blogs/climateqa/stabilize-gg-emissions-effects/sequestration. Based on a graph created from data in the Representative Concentration Pathways Database (Version 2.0.5), http://www.iiasa.ac.at/web-apps/tnt/RcpDb.

The recent Paris Agreement has provided international guidelines for carbon emissions aimed at limiting global temperature increases to 1.5°C above pre-industrial levels. The Intergovernmental Panel on Climate Change (IPCC) has undertaken an extensive study of the approaches necessary to comply (more-or-less) with the Paris Agreement. Figure 2.18 shows the results of such an investigation. An immediate and substantial reduction of carbon emissions is necessary, with a total elimination of emissions by the mid-21st century, followed by an active carbon sequestration program. In addition, 30–50 years of SO_2 injection into the stratosphere is needed. SO_2 counteracts the effects of greenhouse gases in the atmosphere. In this scenario, the mean global temperature increase will overshoot the 1.5°C mark by the mid-21st century, but temperatures will return to the required level by the end of the century. It is interesting to compare this rather aggressive approach to the predicted fossil fuel use in Figure 1.9.

The present chapter has overviewed two significant features of our current energy use: resource depletion and environmental effects. Both features show that our continued dependence on fossil fuels is not sustainable. The next chapter reviews some of the options for energy sources to replace fossil fuels in the near future.

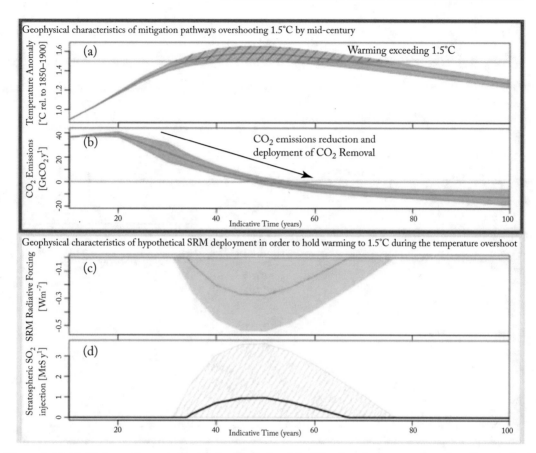

Figure 2.18: Evolution of hypothetical Solar Radiation Modification deployment based on Stratospheric Aerosols Injection in the context of 1.5°C-consistent pathways. (a) Range of global median temperatures as predicted by the Model for the Assessment of Greenhouse Gas Induced Climate Change (Meinshausen, 2011); (b) CO_2 emissions reduction and removal; (c) radiative forcing; and (d) amount of stratospheric SO_2 injection. Based on IPCC. Global Warming Report 1.5°C, http://report.ipcc.ch/sr15/pdf/sr15_chapter4.pdf.

CHAPTER 3

Renewable Energy Sources

3.1 INTRODUCTION

The data presented in the previous chapter demonstrated the need to develop carbon-free energy sources. The time scale for this development on the basis of resource availability is of the order of 50–100 years. However, mitigating the undesirable environmental impact of continued fossil fuel use requires a more proactive approach in dealing with carbon emissions. The magnitude of this problem can be seen by looking at the amount of energy that we obtain from fossil fuels. At present, our annual energy use is about 6.1×10^{20} J. This corresponds to an average power consumption of about 1.9×10^{13} W. About 85% of this, or 1.6×10^{13} W, comes from fossil fuels. The infrastructure that we have developed to produce this quantity of power depends on the high energy density of fossil fuels. For example, to produce this amount of power from coal-fired electric generating stations would require about 16,000 typical facilities (about 1 GW each). By comparison, most sources of carbon-free energy have a much lower energy density. To produce 1.6×10^{13} W from typical large wind turbines (i.e., about 2 MW each operating at 28% capacity) would require about 29 million turbines. If we wanted to replace our fossil fuel use with wind energy on a time scale of 50 years, we would need to construct about 1,600 new 2 MW turbines each day. Certainly, implementing an energy infrastructure that is carbon-free is a formidable task.

In assessing the viability of various non-fossil fuel options, a number of criteria should be considered.

- *Available power.* All sources of energy are limited to some extent. The amount of power available from a given source will determine its maximum possible contribution to our total energy consumption.

- *Longevity.* The length of time a resource will last depends on the total amount of energy available. While we view renewable energy as renewable indefinitely, this may not always be the case.

- *Environmental impact.* While renewable energy production may not directly produce CO_2 emissions, some may contribute to the greenhouse gas balance in the atmosphere as a result of geographical changes that produce greenhouse gases, such as methane, or reduce the natural sequestration of carbon dioxide. Other impacts may include effects on wildlife and direct or indirect effects on human health.

- *Economics.* The cost of renewable energy sources is a factor in determining the time scale for their development. At present, fossil fuels are still plentiful and inexpensive. This fact motivates their continued use and may delay the development of potentially viable, but more expensive, renewable sources.

Additional factors that may influence the development of certain renewable energy sources include social perception, political agendas, security, and safety.

There are a number of options for carbon-free energy and the present chapter concentrates on the five that have made the greatest contribution thus far to carbon-free energy production and are likely to continue to do so in the foreseeable future. These energy sources are: hydroelectric, solar, wind, geothermal, and biofuels. Several other approaches to renewable energy are in the early stages of development and may make important contributions to our energy needs at some point in the future. These approaches include: tidal energy, wave energy, ocean thermal energy conversion, and osmotic energy. For the present discussion, nuclear energy is not considered renewable.

3.2   HYDROELECTRIC ENERGY

Hydroelectric power was one of the first sources of electric power and was initially developed in the late 1870s. It utilizes the gravitational potential of water that results from water level differences on two sides of a dam to turn a turbine-generator assembly to generate electricity. The "head" is defined as the height difference between the water level on the upstream side of the dam and the water level on the downstream side of the dam.

The gravitational potential energy associated with a parcel of water of mass, m, with a head, h, is

$$E = mgh ,$$

(3.1)

where g is the gravitational acceleration, $g = 9.8 \ m/s^2$. The power generated by the turbine is given as

$$P = \frac{dE}{dt} = \frac{dm}{dt} gh ,$$

(3.2)

where dm/dt is the flow rate of the water, in mass per unit time (i.e., kg/s), through the turbine. This may also be expressed as

$$P = \rho \varphi gh ,$$

(3.3)

where ρ is the water density (typically 1,000 kg/m^3) and φ is the flow rate in m^3/s. The conversion of the gravitational potential energy associated with the water to rotational mechanical energy by the turbine and then to electrical energy by the generator, is, of course not 100% efficient. The electrical output may be expressed as

$$P = \eta \rho \varphi g h \,, \tag{3.4}$$

where η is the net turbine-generator efficiency. Efficiencies in the range of 85–90% are typical for large hydroelectric facilities.

Hydroelectric facilities are sometimes categorized as low-head or high-head, depending on the height difference. The term "medium head" is sometimes used to describe facilities with an intermediate head. These categories, however, are not well defined and are sometimes determined by the design of the dam and turbine as well as the actual head.

The general design of a hydroelectric facility is illustrated in Figure 3.1. Water is carried from the upstream reservoir to the turbine and generator through a penstock. The penstock may be inside the dam or, in some facilities, it may be an external pipe as shown in Figure 3.2. There are several different designs of turbines that are used in hydroelectric facilities. For commercial-scale hydroelectric facilities, Kaplan turbines, as shown in Figure 3.3, are most commonly used for low head installations. For high-head installations, Francis turbines, as shown in Figure 3.4, are most commonly used. The hub and blade assembly is referred to as a "runner."

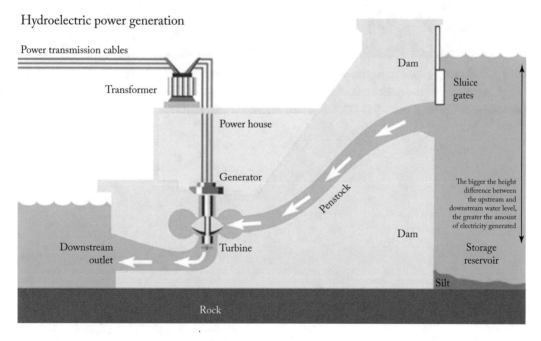

Figure 3.1: **Basic design of a hydroelectric installation.** Based on the first figure from Environment Canada, https://www.ec.gc.ca/eau-water/default.asp?lang=en&n=00EEE0E6-1.

Figure 3.2: Penstock leading to the Lake Margaret Power Station, Lake Margaret, Australia. Image: Steve Lovegrove/Shutterstock.com, https://www.shutterstock.com/image-photo/lake-margaret-power-er-station-australiaoctober-10-542412559?src=MMh5ZVG6Ee0c0JG7gjQaAg-1-72.

Figure 3.3: **A Kaplan turbine runner.** Image: **marekuliasz/Shutterstock.com,** https://www.shutterstock.com/image-photo/old-adjustableblade-kaplan-propellertype-turbine-displayed-761969722?src=d4msN6WcnvmRtrEKo58iLQ-1-2.

Figure 3.4: **A Francis turbine runner.** Image: **Mariusz Hajdarowicz/Shutterstock.com,** https://www.shutterstock.com/image-photo/francis-runer-468329492?src=qonQ0D4zoXv8k3jfATg66A-1-3.

A high-head hydroelectric facility, the Three Gorges Hydroelectric Dam in China, is shown in Figure 3.5. This dam has a head of 181 m and a total capacity of 22,500 MW. At present, it is the largest hydroelectric facility worldwide in terms of total rated capacity. High-head hydroelectric dams block a river and create a reservoir behind the dam to store water. This reservoir serves to minimize variations in hydroelectric output that often result from seasonal differences in river flow rate. The Three Gorges Dam has an associated reservoir that covers an area of 1,084 km^2.

Figure 3.5: The Three Gorges Hydroelectric Dam in Sandouping, Yiling District, Hubei, China. Image: jejim/Shutterstock.com, https://www.shutterstock.com/image-photo/three-gorges-dam-sandouping-yichang-china-128863591?src=5RvWS__e0nTZ-3f0a3RmWA-1-1.

In many cases the reservoir associated with a low-head dam is of minimal size and the design is referred to as a "run-of-the-river" hydroelectric facility. If the total flow rate of the river is large, the hydroelectric output can be substantial despite the small head. This is the case for the 1436 MW La Grande-1 hydroelectric generating station shown in Figure 3.6.

As seen in Chapter 1, hydroelectric energy is the leading form of renewable energy worldwide and, at present, is the largest source of non-fossil fuel energy. Overall, hydroelectric energy accounts for 7% of primary energy worldwide and about 17% of electricity. The major producers of hydroelectric power are given in Table 3.1. In some cases, hydroelectricity is the major source of domestic electricity production, i.e., Norway, Brazil, and Canada, while in other cases, i.e., the U.S., hydroelectricity is a minor component of electricity production.

Figure 3.6: La Grande-1 hydroelectric station on the La Grande Rivière in Québec, Canada. This dam is part of Hydro-Québec's projet de la Baie-James. (CCBYSA3.0, https://creativecommons.org/licenses/by-sa/3.0/deed.en, https://commons.wikimedia.org/wiki/Commons:GNU_Free_Documentation_License,_version_1.2), https://en.wikipedia.org/wiki/La_Grande-1_generating_station#/media/File:LG1.jpg.

Table 3.1: The largest producers of hydroelectric power (>10^{14} Wh/y as of 2014). Data are from https://en.wikipedia.org/wiki/Hydroelectricity

Country	Annual Hydroelectric Production (TWh)	Installed Capacity (GW)	Capacity Factor	Fraction of Domestic Electricity Production
China	1064	311	0.37	0.187
Canada	383	76	0.59	0.583
Brazil	373	89	0.56	0.632
United States	282	102	0.42	0.065
Russia	177	51	0.42	0.167
India	132	40	0.43	0.102
Norway	129	31	0.49	0.960

While infrastructure costs of hydroelectric power can be considerable, operational costs are minimal, and the resulting electricity cost per kWh is relatively low. Although the output may have seasonal fluctuations, the short-term variations are generally minimal and electricity generation is more consistent than for most other renewable sources, such as wind or solar.

Although hydroelectric power does not directly produce carbon dioxide emissions, it is not, in all cases, without adverse environmental consequences. These effects are most notable for high head hydroelectric installations. The large area covered by the upstream reservoir can contribute to greenhouse gases in two ways. The replacement of land that is potentially covered by vegetation with water reduces the amount of CO_2 that is sequestered. As well, vegetation in areas that are flooded by the construction of a high-head hydroelectric dam will decay and produce CO_2 and methane. These problems can be significant in tropical regions but are much less serious in temperate regions.

Since hydroelectricity is, compared to other renewable sources of energy, a very mature technology, much of the available resources worldwide have already been utilized. This is particularly true for high-head resources in the industrialized world. Undeveloped resources may be viewed as either technically feasible or both technically feasible and economically viable. At present, the development of new technically feasible and economically viable resources might increase overall hydroelectric production by about 50% or so. At some point in the future hydroelectricity may increase by a factor of 2 or 3, leading to a total contribution of 15–20% of total world primary energy production. Much of potential future development is likely to occur in less industrialized countries and may be primarily low-head facilities.

Although hydroelectricity is often viewed as a renewable energy source, it is not necessarily renewable indefinitely. This is particularly the case for some high-head installations. Silt carried downstream may accumulate behind a dam. Reduction in power output and increases in maintenance costs may, at some point, make the facility no longer economically viable.

Additional considerations that apply mainly to high-head hydroelectric facilities include the following.

- Relocation of residents in areas that are flooded by the creation of a reservoir. As well, cultural or historical sites may be destroyed, and cemeteries may need to be relocated. Over 1 million residents had to be relocated during the construction of the Three Gorges Hydroelectric Dam.

- Although hydroelectric power is generally considered to be safe, catastrophic dam failures may be responsible for significant fatalities. The most notable event of this type was the failure of several dams on the Ru River in China in 1975 as a result of raising water levels caused by Typhoon Nina. There were approximately 26,000 deaths directly attributed to the resulting flooding and an estimated 145,000 additional deaths from famine and related health issues.

- Hydroelectric dams are often designed for the purpose of water control, as well as the production of electricity. However, in many regions the accumulation of silt behind the dam not only reduces the effectiveness of the hydroelectric facility but reduces the transport of nutrients downstream to agricultural areas.

- Finally, hydroelectric dams may also have ecological effects. The replacement of forests or grasslands with a reservoir certainly effects the diversity of wildlife in the area. These effects may be either positive or negative. In general, dams have a negative effect on fish mobility.

3.3   WIND ENERGY

Wind energy has been used since antiquity to propel sailing vessels. More than 500 years ago wind became a common source of energy in Europe for grinding grain. More recently it was used extensively to pump water for agricultural purposes. In the 20th century, wind turbines were developed for the purpose of generating electricity.

The energy available from the wind can be determined by the following simple analysis. A parcel of air of mass, m, moving with a velocity, v, has kinetic energy given by

$$E = \frac{1}{2}mv^2 .$$ (3.5)

If the parcel has a cross sectional area, A, normal to the direction of the wind and a length, L, parallel to the direction of the wind, then the mass is given in terms of the air density, ρ, as

$$m = \rho LA .$$ (3.6)

Substituting into Equation (3.5) gives

$$E = \frac{1}{2}\rho LAv^2 .$$ (3.7)

The power, which is energy per unit time, is

$$P = \frac{1}{2}\rho A \frac{dL}{dt}v^2 = \frac{1}{2}\rho Av^3 .$$ (3.8)

Equation (3.8) gives the total power available from the wind over an area, A. Power may be extracted from the wind by a wind turbine (see Figure 3.7). These are examples of the typical three-blade horizontal axis wind turbines in use today. Other turbine geometries have been developed, as in Figure 3.8, but have not proved to be as practical as the three-blade horizontal axis design.

Figure 3.7: Two Suzlon S97 2.1 MW wind turbines. Image courtesy of Richard A. Dunlap.

Figure 3.8: The Éole Darrieus wind turbine near Cap-Chat, Gaspésie, Québec, Canada. At 64 m in diameter and 96 m high, this 4 MW turbine is the world's largest vertical axis wind turbine. It was constructed in 1987 and is no longer operational. Image: meunierd/Shutterstock.com, https://www.shutterstock.com/image-photo/cap-chat-quebeccanada-august-25-2012-1156762006.

In order to extract power from the wind, the wind velocity must be reduced. Obviously, in order to extract all of the power available from a moving parcel of air, the air must be stopped. If a wind turbine was designed to stop the wind, the air would tend to flow around the turbine, thereby reducing its efficiency. The maximum theoretical efficiency of a wind turbine is 59% and this is referred to as the Betz limit. Thus, we can write the total power available for a turbine with a rotor area, A, as

$$P = \frac{1}{2} \eta \rho A v^3 \, .$$
(3.9)

where η is the efficiency.

The actual efficiency of a wind turbine is determined by the turbine design and the velocity of the wind. This relationship is illustrated in Figure 3.9 for the common three-blade turbine. The efficiency is plotted as a function of the tip speed ratio, which is the ratio of the velocity of the tip of the rotor divided by the velocity of the wind. It is seen in the figure that a maximum efficiency of close to 50% can be achieved for a tip speed ratio of around 7.

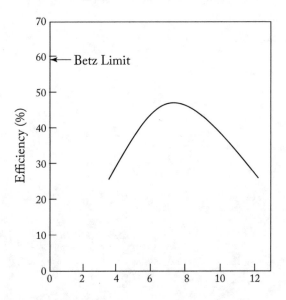

Figure 3.9: Efficiency of a three-blade horizontal axis turbine as a function of the tip speed ratio.

The connection of a wind turbine to the electric grid is not straightforward. In general, one must consider three important factors: the voltage, the frequency, and the phase. All three must be compatible with the power on the grid. The exact details of the interface between the electricidal out put of the generator associated with a wind turbine depends, to a large extent, on the details of the generator design. However, a basic schematic of a typical approach to this problem is illustrated in Figure 3.10. It is common in such systems to convert the AC output from the generator to DC by means of a rectifier. This DC voltage is then converted back to AC at the proper frequency and phase as shown in the figure. The AC voltage is then adjusted to the grid voltage by a transformer.

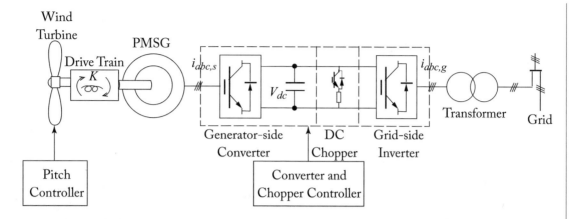

Figure 3.10: Schematic design of a grid connection for a wind turbine (Figure 1 from Wu et al., 2013).

The dependence of the turbine power on the cube of the wind velocity means that it is important to locate wind turbines in areas which have high wind velocities. Obviously, the wind velocity at any location does not remain constant but has both daily and seasonal variations. A typical distribution for measured wind speeds is illustrated in Figure 3.11. Wind speed distributions are generally well approximated by the Rayleigh distribution of the form

$$f(v) = \frac{v}{\sigma^2} \exp\left[-\frac{v^2}{2\sigma^2}\right], \tag{3.10}$$

where σ is a fitting parameter. Wind turbines have a range of operational rotational speeds from a minimum (the cut-in speed) to a maximum safe rotational speed (the cut-out speed) where the turbine is shut down and the rotors locked to prevent damage to the unit. It is important to optimize the design of the turbine for a particular location to best make use of the distribution of wind speeds. Typical installations have a capacity factor of around 30%, that is, the ratio of the total annual output to the output for the turbine running continuously at full capacity.

In recent years, it has become increasingly common to place wind turbines offshore in coastal regions. As average wind speed typically increases with distance from shore, energy produced also increases with distance from shore. Placing turbines offshore also has the advantage of not interfering with potential land use and may (or may not) reduce the social impact of the facility.

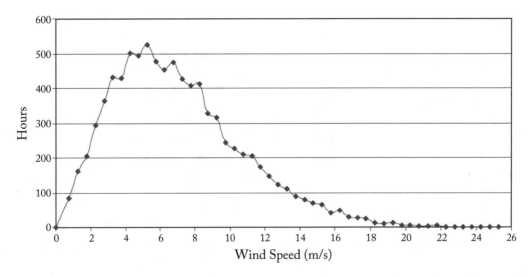

Figure 3.11: Frequency distribution of wind speeds at a height of 40 m at the Lee Ranch Wind Monitoring Station in the San Mateo Mountains of New Mexico. Data are for the year 2002. Based on https://windpower.sandia.gov/other/LeeRanchData-2002.pdf.

The placement of turbines in a wind farm is also of significant importance. Placing turbines too far apart will reduce the number of turbines that can be accommodated in a certain area. However, placing turbines too closely together may reduce the total energy output as one turbine may block the wind from reaching other turbines. Roughly, spacing turbines 8–12 rotor diameters apart in the prevailing wind direction and 2–4 rotor diameters apart in the orthogonal direction has usually been considered appropriate. Thus, for turbines with rotor diameter, R, the land area per turbine would be $16R^2$ to $48R^2$. Since the area of the rotor goes as πR^2 and the energy produced by the turbine will scale as the rotor area, then it might seem that for a given land area, a lot of small turbines would be equivalent to a smaller number of large turbines in terms of total energy production. This, however, is not true; a smaller number of large turbines produces more energy. This results from the height dependence of the wind velocity. This height dependence is generally modeled using the expression

$$v(h) = v(h_0) \frac{\ln(h/\delta)}{\ln(h_0/\delta)} .$$ (3.11)

where h_0 is a reference height, $v(h_0)$ is the velocity at height h, and δ is a measure of the surface features of the land below the turbine. Typical results for this approach are illustrated in Figure 3.12

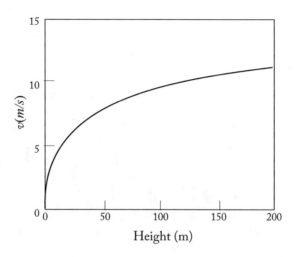

Figure 3.12: Typical height dependence of wind velocity as predicted by Equation (3.11) for roughness class 3 and a velocity of 10 m/s at 125 m. Roughness class 3 is appropriate for turbines surrounded by forest.

The use of wind energy has grown substantially in the past twenty years. This is illustrated by the trends in the total worldwide installed capacity as shown in Figure 3.13. The installed capacity and total annual electricity production (as of 2015) for the top wind power-producing countries is summarized in Table 3.2. The total electricity generated by wind in 2015 was 834 TWh or 3.0×10^{18} J (3.0 EJ). This amounts to less than 1% or total global primary energy production. The total viable wind energy worldwide is estimated to be around 2,000 EJ/y.

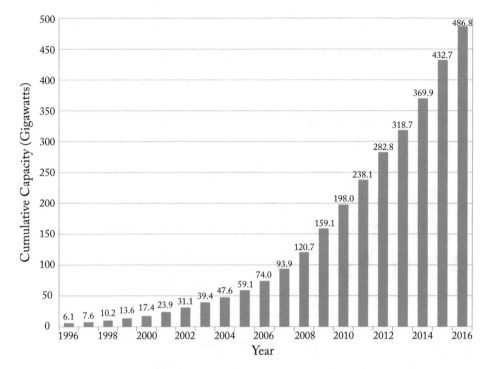

Figure 3.13: Growth in total world installed wind capacity since 1996. Based on https://commons.wikimedia.org/wiki/File:Global_Wind_Power_Cumulative_Capacity.svg.

The commitment of various nations to the development of wind energy is best assessed by a comparison of the fraction of electricity generated from wind. This is summarized in Table 3.3. It is clear that Denmark, as well as several other European countries, has taken the utilization of wind energy very seriously.

Table 3.2: Installed capacity and electricity generated for the top wind power producing countries in 2015

Country	Installed Capacity (GW)	Annual Production (in TWh)
China	145.1	185.8
United States	72.6	190.7
Germany	44.9	78.9
India	25.1	42.8
Spain	23.0	48.1
United Kingdom	13.9	40.3
Canada	11.2	26.2
France	10.4	21.2

Table 3.3: Countries with the largest wind production as a percentage of total electricity generation

Country	Year	Percent Total Electricity
Denmark	2015	42.1
Portugal	2013	23
Spain	2011	16
Ireland	2012	16
United Kingdom	2015	11
Germany	2011	8
United States	2016	5.5

Environmental consequences of wind turbines have been the subject of considerable discussion in recent years. Typical concerns include noise and effects on wildlife (particularly birds).

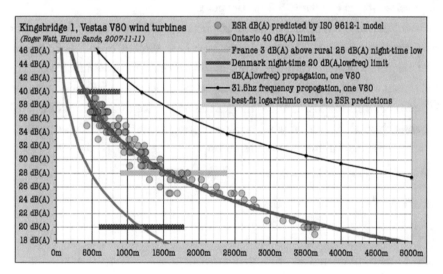

Figure 3.14: Typical noise levels as a function of distance from a Vestas V80 wind turbine. Reproduced with permission of Roger W. Watt. http://retirees.uwaterloo.ca/~rwwatt/esr.png.

Noise levels for a typical wind turbine are shown in Figure 3.14. In many jurisdictions wind turbines are placed a minimum of 300 m from residential buildings. The graph shows that, at this distance, noise levels from the turbine are around 40–42 dB (decibels). By comparison, noise from a typical home refrigerator is around 40 dB and from a window air conditioner is around 50 dB. Typical background noise levels in residential areas is between 40 and 45 dB, meaning that, at 300 m, a wind turbine will probably not be detected above ambient noise levels. In very quiet rural areas, background noise levels might be around 30 dB. Figure 3.14 shows that this corresponds to a typical wind turbine noise level at a distance of about 1.5 km. It has been speculated that very low-frequency sound (i.e., infrasound) at less than about 20 Hz which is not audible to the human ear, may have adverse health related consequences, such as sleep-related disorders, concentration difficulties, and headaches. Clear correlations between infrasound and health issues have not been established and further research is needed in this area.

Avian mortality has often been used as an argument against wind energy development. A 2013 study by Environment Canada concluded an average mortality rate of 8.2 birds per year per wind turbine. A similar Spanish study that examined wind turbine related bird deaths between 2005 and 2008 found an average of 1.3 deaths per turbine per year. A recent study has put bird mortality from wind turbines in the context of deaths caused by other energy production technologies (Sovacool, 2009). Fossil fuel-based electricity generation produces bird deaths as a result of mining operations, bird collisions with equipment, pollution, and global warming. Wind-generated electricity is responsible for an estimated 0.3 deaths per GWh electricity generated, while fossil fuels are responsible for an estimated 5.2 deaths per GWh.

Human injuries and fatalities are also an important consideration for any energy generation technology. Since about 1980, 164 human deaths have been directly or indirectly attributed to the generation of electricity by wind turbines. The majority of these fatalities involved either the manufacture and transportation of turbine components or the construction or maintenance of turbines.

Wind turbines can also pose a risk to the general public. Ice thrown from turbine blades presents one potential hazard. Failure of wind turbines during operation can also constitute a hazard. In general, turbine failures fall into three categories:

1. blade failure (Figure 3.15(a)),

2. nacelle fire (Figure 3.15(b)), and

3. Ttower failure (Figure 3.15(c)).

Often failure of one component can precipitate other failures.

(a) Image: Carol Heesen/Shutterstock.com, https://www.shutterstock.com/image-photo/one-broken-wing-on-single-wind-33225310?src=lGFzpn_d3NhtMvo1vxWi0Q-1-4

(b) Image: Dmitri Ma/Shutterstock.com, https://www.shutterstock.com/image-photo/view-burned-wind-turbine-over-blue-276604328?src=lGFzpn_d3NhtMvo1vxWi0Q-1-0

(c) Image: juerginho/Shutterstock.com, https://www.shutterstock.com/image-photo/by-storm-toppled-wind-wheel-near-787831234?src=lGFzpn_d3NhtMvo1vxWi0Q-1-50.

Figure 3.15: Modes of wind turbine failure: (a) blade failure, (b) nacelle fire, and (c) tower failure.

3.4   SOLAR THERMAL ENERGY

Solar energy is the single largest renewable energy resource. From Chapter 2 it was seen that the average insolation outside the atmosphere is 342 W/m². Slightly less than half of this, or about 168 W/m², is absorbed by the earth's surface, the remainder being absorbed in the atmosphere or reflected back into space. Average insolation varies considerably at different locations on earth. Latitude plays an important role with the greatest insolation in equatorial regions and minimal insolation near the poles. Local climate is also important, as insolation is reduced in areas with frequent cloud cover. The earth has a surface area of 5.1×10^{14} m². About 29% of this, or 1.49×10^{14} m², is land. The total insolation on the land area of the earth is, therefore, about 2.5×10^{16} W. This is more than 1,500 times the current power consumption of society (1.6×10^{13} W). Thus, the utilization of less than 0.1% of the total solar radiation incident on the land surface of the earth would be needed to fulfill all our energy needs.

Solar radiation may be used in two ways: (1) the direct use of heat produced by the absorption of the radiation, as discussed in the present section; and (2) the conversion of solar radiation into electricity, as discussed in the next section.

Heat produced by the absorption of solar radiation is best used locally to provide space heating and domestic hot water. Passive solar systems based on efficient building design and location can provide a substantial contribution to space heating needs. Active systems, where heat is transported by a carrier fluid, are more versatile. There are several possible designs for active solar collectors. Four of these—flat plate collectors, evacuated tube collectors, transpired collectors, and concentrating solar collectors—are discussed below.

3.4.1   FLAT PLATE COLLECTORS

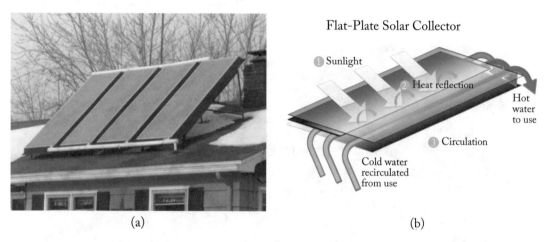

Figure 3.16: (a) Photograph of residential flat plate solar collectors for water and space heating. Image courtesy of Richard A. Dunlap. (b) Schematic of the basic design of a flat plate solar collector. Based on https://www.epa.gov/rhc/solar-heating-and-cooling-technologies. Source: EPA.

Flat plate solar collectors are probably the most common design for residential use. The flat plate collector consists of a box containing a backing plate (painted black) that absorbs solar radiation. A pipe carrying a working fluid is connected to the backing plate so that the heat is transferred to the fluid. Collectors may be unglazed (open on the front) or glazed (covered with a transparent sheet; usually glass). The glazed collectors, as shown in Figure 3.16, are most common and are more efficient than unglazed collectors. In the glazed collector, the glass allows the short wavelength solar radiation to pass into the box but blocks the long wavelength radiation emitted by the heated backing plate from escaping. This greenhouse effect traps heat inside the collector and this facilitates the transfer of thermal energy to the working fluid. Efficiencies of around 50% are typical for glazed collectors. Flat plate collectors may utilize water as the working fluid to provide hot water for space heating or domestic hot water use or they may use a working fluid such as water containing antifreeze to transfer heat to water through a heat exchanger.

3.4.2   EVACUATED TUBE COLLECTORS

Evacuated tube collectors are a more recent development than flat plate collectors that, in some cases, provide improved performance. An evacuated tube collector is shown in Figure 3.17(a) along with a diagram of its basic design. The collector consists of an array of parallel evacuated glass tubes. Each tube contains a metal tube (painted black to absorb solar radiation) that is closed on both ends, as shown in Figure 3.17(b). The tubes contain a fluid (usually a mixture of water and propylene glycol) that undergoes a phase transition from the liquid state to the vapor state. As the solar radiation heats the fluid and it evaporates, the hot vapor rises to the top of the tube and transfers heat to a pipe containing flowing water. The heated water can be utilized for space heating or hot water, in the case of the flat plate collector.

(a) Image: Beautiful landscape/Shutterstock.com, https://www.shutterstock.com/image-photo/vacuum-collectors-solar-water-heating-system-243692335?src=GEvzTzUT-rC2nIZtPoExWQ-1-2.

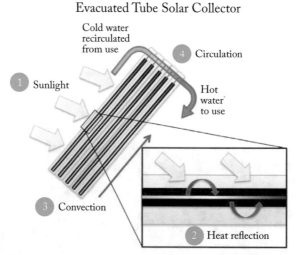

(b) Source: EPA https://www.epa.gov/rhc/solar-heating-and-cooling-technologies.

Figure 3.17: (a) Photograph of evacuated tube solar collectors and (b) schematic of the basic design of an evacuated tube solar collector.

3.4.3   TRANSPIRED SOLAR COLLECTORS

The design of a transpired solar collector, sometimes referred to as a "solar wall" is shown in Figure 3.18(a). This basically consists of an outer wall made of black sheet metal with an air space separating it from the main building wall. Solar radiation is absorbed by the black metal and heats the air in the space. A fan-driven ventilation system, as shown in Figure 3.18(b), draws the hot air into the building and integrates it into the building heating system. Such systems are most commonly used for commercial and industrial buildings rather than residences.

(a) Conserval Engineering Inc./Wikimedia.com/CC-BY-SA-3.0, https://creativecommons.org/licenses/by-sa/3.0/deed.en, https://commons.wikimedia.org/wiki/File:GTAA_-_SolarWall.jpg.

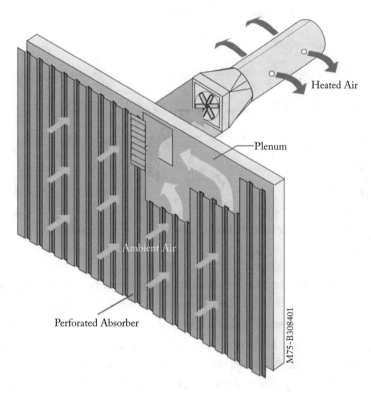

Heated Air

Plenum

Ambient Air

Perforated Absorber

M75-B308401

(b) Based on **NREL**, https://www.nrel.gov/docs/fy06osti/29913.pdf.

Figure 3.18: (a) Transpired solar collector on the facade of the Fire and Emergency Services Training Institute (FESTI), at the Toronto Pearson International Airport, Ontario, Canada, and (b) basic design of a transpired solar collector.

3.4.4   CONCENTRATING SOLAR COLLECTORS

Concentrating solar collectors use a reflector to concentrate solar radiation from a large area onto a small area. A typical example is illustrated in Figure 3.19, where a parabolic reflector concentrates radiation onto a pipe carrying water. Because solar energy from a large area is used to heat water in a small volume, high temperatures and/or large flow rates can be achieved. These systems are best used for industrial, commercial, or agricultural applications. This type of geometry is discussed further in the next section on the generation of electricity from solar radiation.

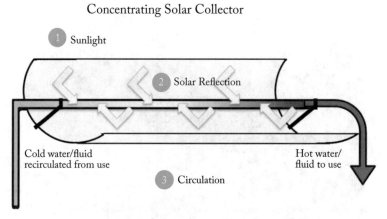

Figure 3.19: **A concentrating solar collector. Based on** https://www.epa.gov/rhc/solar-heating-and-cooling-technologies. **Source: EPA.**

As indicated above, the amount of available solar energy is basically unlimited (compared with our needs) and is renewable indefinitely. The direct utilization of thermal energy derived from solar radiation is, however, generally limited to local installations, in most cases for residential applications for space heating and hot water. In general, the implementation of solar thermal energy has few, if any, adverse effects. Although the lifetime of equipment may be limited, maintenance costs are typically minimal. Operational cost, i.e., electricity for pumps or blowers to circulate the working fluid or air, respectively, are generally small compared to the energy acquired. From an economic standpoint, individual users need to assess infrastructure and maintenance/operational costs in terms of energy savings to determine if the resulting payback period is acceptable.

3.5   SOLAR THERMAL ELECTRICITY

Although electricity produced from solar energy may, in some cases, be used in stand-alone residential systems, its main contribution to renewable energy is in the form of large-scale, grid-connected facilities. As discussed below, production of electricity from solar energy takes one of two basic approaches: thermal generation or photovoltaics.

Most commonly the generation of electricity from solar thermal energy utilizes concentrating solar collectors, although experimental facilities using some other approaches have been constructed, including, solar ponds and solar updraft towers. The discussion here will deal with concentrating collectors.

Following from Figure 3.19, Figure 3.20(a) shows one approach to generating electricity from solar thermal energy where reflective parabolic troughs focus sunlight onto to fluid carrying pipes. A working fluid is heated and used to generate steam to turn a turbine and generator. Figure 3.20(b) shows a parabolic dish system where concentrated solar radiation can be used to heat a working fluid, as in the case of the parabolic trough, or can be converted directly into mechanical energy to drive a generator by means of a Stirling engine.

(a)

(b)

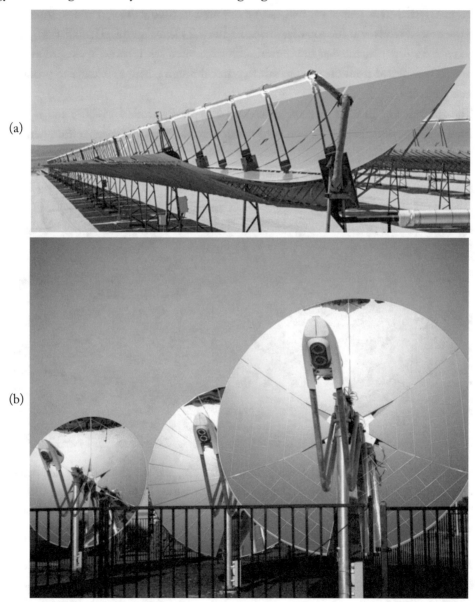

Figure 3.20: (a) Parabolic trough collectors which focus solar radiation on a pipe carrying a working fluid. Tom Grundy/Shutterstock.com, https://www.shutterstock.com/image-photo/solar-electric-pow-er-plant-parabolic-mirrors-191334824?src=Yyz3auMGLk4EDmKSOWML4w-1-43. (b) An array of parabolic dish collectors with Stirling engines. J.D.S/Shutterstock.com, https://www.shutterstock.com/image-photo/mirrored-parabolic-dish-solar-energy-equipment-57711145.

Another approach to the construction of a concentrating solar collector is the central receiver geometry, sometimes called a "central tower" or "power tower." Sunlight is reflected by an array of flat mirrors, referred to as heliostats, that can be oriented to follow the sun and to direct reflected light to the central receiver at the top of a tower. Figure 3.21 show a typical installation. Because of the high temperatures generated at the central receiver, molten salt is sometimes used as a working fluid. Heat is transferred from the molten salt to water through a heat exchanger to produce steam to operate a turbine.

The efficiency of all solar thermal generation is typically around 15–20% and is limited by the Carnot efficiency of the heat engine used to convert heat into the mechanical energy needed to drive the generator.

Figure 3.21: Central receiver ("power tower") solar-generating station. The array of heliostats can be seen in the image and these heliostats can be angled to reflect the sun to the receiver at the top of the tower when the station is in operation. StockStudio/Shutterstock.com, https://www.shutterstock.com/image-photo/solar-power-plant-mirrors-that-focus-1025906944.

A summary of the largest operating solar thermal power stations (≥ 150 MW capacity) is given in Table 3.4. It is seen that the largest capacity station, a central receiver system is the Ivanpah Solar Power Facility in California. The remainder of major facilities worldwide are parabolic trough systems. Table 3.5 summarizes the capacity worldwide by country. The world total from these data is 4,815 MW capacity.

Table 3.4: World's largest solar thermal-generating stations. Capacity is maximum rated electrical output

Name	Type	Capacity (MW)	Location
Ivanpah Solar Power Facility	Central receiver	392	San Bernardino County, CA
Solar Energy Generating Systems	Parabolic trough	361	Mojave Desert, CA
Ouarzazate Solar Power Station	Parabolic trough	360	Ghassate, Spain
Mojave Solar Project	Parabolic trough	280	Barstow, CA
Solana Generating Station	Parabolic trough	280	Gila Bend, AZ
Genesis Solar Energy Project	Parabolic trough	280	Blythe, CA
Solaben Solar Power Station	Parabolic trough	200	Logrosán, Spain
Solnova Solar Power Station	Parabolic trough	150	Sanlúcar la Mayor, Spain
Andasol Solar Power Station	Parabolic trough	150	Guadix, Spain
Extresol Solar Power Station	Parabolic trough	150	Torre de Miguel Sesmero, Spain

Table 3.5: Total solar thermal-generating capacity by country

Country	Total (MW)
Spain	2,300
United States	1,738
India	225
South Africa	200
Morocco	180
United Arab Republic	100
Algeria	25
Egypt	20
Australia	12
China	10
Thailand	5

3.6   PHOTOVOLTAICS

Photovoltaic cells utilize the properties of semiconducting junctions to convert sunlight directly into an electric current. The most common semiconducting material used in photovoltaic cells is silicon. Silicon has a diamond crystal structure and each silicon atom is tetrahedrally coordinated to its neighboring silicon atoms as shown in Figure 3.22. Elemental silicon has a valence of 4, so that each covalent bond between neighboring silicon atoms consists of two shared electrons, one from each atom.

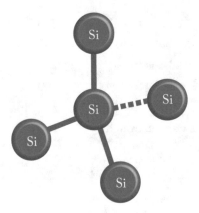

Figure 3.22: Tetrahedral coordination in pure silicon.

If a silicon atom is replaced by an atom with a valence of 5 (e.g., phosphorus) then one extra valence electron will be present, and this electron can move freely within the structure and can readily contribute to electrical conduction. This substitution process is referred to as doping and, in this case, the impurity atoms are referred to as donor atoms because they donate an extra electron. Such a material is referred to as an n-type semiconductor (n for negative, indicating the sign of the charge of the donated electron). Analogously, substituting a valence 3 atom (e.g., boron) into silicon means that one of the bonding electrons is missing. This missing electron is referred to as a hole and behaves as an effective positive charge. Such a material is referred to as a p-type semiconductor (p for the positive charge of the hole) and the impurity atoms are referred to as acceptors.

Semiconductor devices are constructed using various combinations of n-type and p-type semiconducting materials. The simple p-n junction, which functions as a diode, is illustrated in Figure 3.23. Electron carriers in the n-type region are repelled away from the junction by acceptor atoms (which have gained an electron and have become negatively charged ions) in the p-type material. Similarly, the hole carriers in the p-type material are repelled away from the junction by the ionized donor ions in the n-type material. This situation leads to a distribution of the charges in the material as shown in Figure 3.23. The n-type region is electrically neutral, as the charge of the free electrons is compensated by an equal number of the donor atoms which have been ionized and have a net positive charge. Similarly, in the p-type region the charge of the free holes is compensated by an equal number of acceptor atoms that have picked up an extra electron giving them a net negative charge. The region around the junction is mostly free from mobile electron or hole carriers and is referred to as the depletion region. In this region the donor atoms and acceptor atoms all carry positive or negative charges, respectively, leading to the establishment of an electric field across the junction pointing from left to right in the figure.

d+	d	d+	d+	d+	a-	a-	a-	a	a
	e						h		
d	d+	d+	d+	d+	a-	a-	a-	a- h	a-
e									
d+	e d	d+	d+	d+	a-	a-	a-	a	a-
								h	
	e						h		
d+	d	d+	d+	d+	a-	a-	a-	a-	a

n-type region	Depletion region	p-type region

Figure 3.23: The p-n junction.

Now we can understand what happens in the semiconducting junction when light is incident upon it. Light consists of photons, each with an energy, E, that is related to its frequency, v, as

$$E = hv , \qquad (3.12)$$

or in terms of wavelength, λ,

$$E = \frac{hc}{\lambda} . \qquad (3.13)$$

where c is the speed of light. The photon will interact with the charges in the semiconducting material, and if it has sufficient energy, then it can ionize one of the atoms and, thereby, produce an electron-hole pair. The minimum energy required for this to happen in referred to as the "energy gap." When an electron-hole pair is formed, then the electron will travel to the left (in Figure 3.23) and the hole will travel to the right. Both of these flows of charge represent a positive current flowing from left to right in the figure and this net current can do work on an external circuit.

The production of a current across a p-n junction in response to incident photons forms the basis for the operation of a photovoltaic cell. In a practical photovoltaic cell, it is important to maximize the active area and to ensure that photons can effectively reach the junction. Figure 3.24 shows the basic design of a simple photovoltaic cell. To optimize the quantity of light reaching the junction, the front electrode is made from thin wires or a fine metallic mesh and the n-type semiconducting layer is very thin (~1 μm). In practical devices, more complex geometries involving multiple junctions are sometimes used.

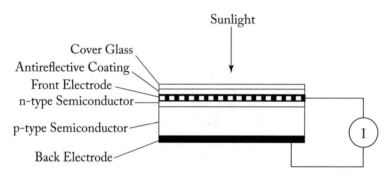

Figure 3.24: Basic design of a simple photovoltaic cell.

There are a variety of photovoltaic cell designs using various semiconducting materials. There is typically a trade-off between efficiency and cost. More expensive cells may utilize more complex and expensive fabrication methods and/or more expensive (and often less available) materials. Commercial cells most frequently use silicon-based semiconducting materials and have efficiencies in the range of 15–20%, where the efficiency is the ratio of the electrical power output to the total incident solar power. Experimental cells can have efficiencies up to around 45%.

Photovoltaic applications range from small (mW) stand-alone devices such as watches and calculators to medium-size off-grid devices (W to kW) such as power for remote instrumentation (e.g., radio transmitters, as shown in Figure 3.25) to multi-MW grid connected facilities, as shown in Figure 3.26.

In recent years, grid-connected photovoltaic capacity has grown significantly. Figure 3.27 shows that, for the past 25 years or so, the growth has been roughly exponential (linear on a semi-logarithmic plot as a function of time). Table 3.6 summarizes the capacity and fraction of electricity consumption from photovoltaics in the ten leading countries. In terms of the fraction of electricity consumption, European Union countries have been among the most active in developing solar resources.

Figure 3.25: A close up of a photovoltaic cell array used to power a remote radio transmitter. Image courtesy of Richard A. Dunlap.

Figure 3.26: Aerial view of Longyangxia Dam Solar Park in China. This photovoltaic facility has a rated capacity of 850 MW and covers an area of 23 km^2. Image: Shutterstock.com, https://www.shutterstock. com/image-photo/longyangxia-dam-solar-park-china-largest-1194149692?src=IzUys2M-07u5hIhp-FEHlpg-1-34.

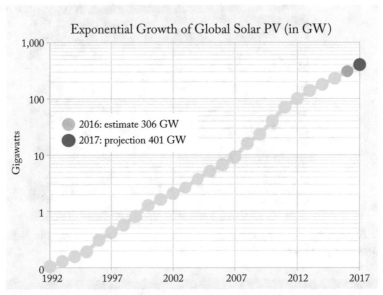

Figure 3.27: Total worldwide installed photovoltaic capacity from 1992–2017. Based on https://commons.wikimedia.org/wiki/File:PV_cume_semi_log_chart_2014_estimate.svg.

Table 3.6: Solar photovoltaic capacity and fraction of total electricity consumption for the top ten nations (as of 2017)

Country	Installed Capacity (GW)	% Total Electricity Consumption
China	131	1.8
United States	51	1.4
Japan	49	4.9
Germany	42	6.7
Italy	19.7	7.5
India	18.3	2.2
United Kingdom	12.7	3.4
France	8	1.6
Australia	7.2	2.4
Spain	5.6	3.2

It is interesting to compare the data in Tables 3.5 and 3.6. It is clear that solar thermal electricity generation is only a small fraction of that from photovoltaics, actually about 1%. While future developments in solar thermal generation may increase the utilization of this approach, photovoltaics are generally simpler and more reliable and will likely continue to account for the vast majority of solar electricity.

While it might appear that solar energy would be fairly benign from an environmental stand-point, this is not entirely the case. The environmental impact of solar energy stems from the low-energy density of solar radiation and the extensive manufacturing requirements for photovoltaic cells. The land required for photovoltaic facilities generally has limited other uses. As the photovoltaic arrays block incident sunlight, the land beneath the collectors is not suitable for agricultural use. It may be possible to place collectors on building surfaces, above parking lots or to cover road surfaces with photovoltaic cells. However, there has not yet been widespread use of these approaches for grid-scale photovoltaic facilities.

In the past, the cost of electricity produced using photovoltaic cells was higher than for many other methods. In recent years, developments in fabrication technology has reduced the cost appreciably. Figure 3.28 shows the historical cost for photovoltaic electricity and U.S. Department of Energy goals for future cost reduction.

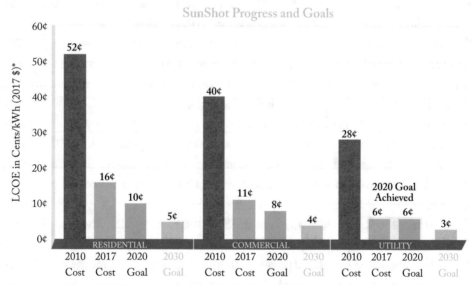

*Levelized cost of electricity (LCOE) progress and targets are calculated based on average U.S. climate and without the ITC or state/local incentives. The residential and commercial goals have been adjusted for inflation from 2010–17.

Figure 3.28: Cost of electricity (in US¢/kWh) produced by photovoltaics for different applications. Based on https://www.energy.gov/eere/solar/photovoltaics.

3.7   GEOTHERMAL ENERGY

Geothermal energy has been used since ancient times as a source of hot water for bathing. In the 14th century the city of Chaudes-Aigues, France constructed a district heating system utilizing geothermal energy. The first commercial use of geothermal energy to generate electricity occurred in Italy in 1911. The availability of geothermal energy is a consequence of the fact that the interior of the earth is hot. The earth"s internal heat is the result of three independent factors:

1. the heat that remains from the thermal energy of the material that condensed to form the earth during the formation of the solar system;

2. heat generated by the friction in the liquid outer core of the earth caused by tidal forces due to the moon and sun; and

3. heat produced by the decay of radioactive nuclei (i.e., potassium, thorium, and uranium) inside the earth's core.

While the exact contribution of each of these three factors to the heat inside the earth is not known, it is generally believed that the third of these (radioactive decay) dominates.

The average net heat flux flowing outward from the earth's surface amounts to 0.087 W/m^2. This is small compared to the net solar energy flux incident on the surface of the earth of 168 W/m^2. Nevertheless, this geothermal heat can provide a useful contribution to our thermal and/or electrical energy needs. The average thermal gradient below the surface of the earth is about 30°C/km. While the natural heat capacity of the earth is often utilized in ground source heat pumps for residential use, it is regions of the earth where there is an enhanced thermal gradient (>~40°C/km) that are most useful for large scale use of geothermal heat and for the generation of geothermal electricity. The most useful geothermal deposits are at temperatures above 100°C and contain pressurized hot water and/or steam.

Geothermal deposits are primarily associated with tectonic plates. These are illustrated in Figure 3.29. Figure 3.30 shows how geothermally active areas are formed at the edges of tectonic plates during the subduction of oceanic plates by continental plates and near the center of oceanic plates during the formation of mid-oceanic ridges. The former case occurs, for example, along the west coast of North and South America and along the east coast of Asia. The latter situation occurs, for example, along the Mid-Pacific ridge, which passes through Hawaii, and the Mid-Atlantic ridge, which passes through Iceland. It is in these active regions that geothermal energy resources are most plentiful. It is also in these active regions that volcanic activity is most prevalent.

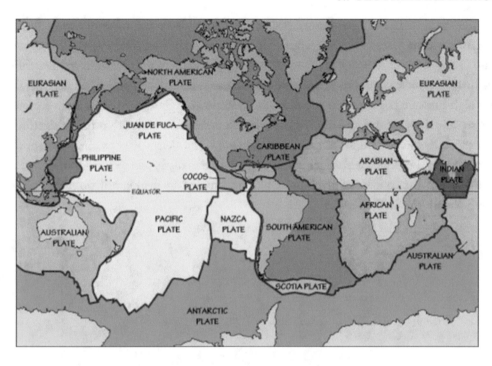

Figure 3.29: **Map of the world's tectonic plates. Source: USGS,** https://geomaps.wr.usgs.gov/parks/pltec/pltec2.html.

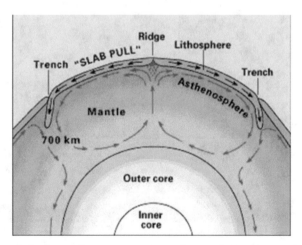

Figure 3.30: Formation of geothermally active areas at edges and the center of tectonic plates. Source: USGS, https://pubs.usgs.gov/gip/dynamic/unanswered.html.

The actual design of a geothermal power plant depends on a number of factors such as the temperature and pressure of the deposit, the fractions of water and steam in the deposit and the end use of the geothermal energy. Figure 3.31 shows a diagram of a flash steam geothermal power plant. This is the most common type of geothermal plant and utilizes geothermal deposits that are largely high-pressure hot water. As the diagram illustrates, the geothermal fluid from a production well is separated into steam, which is used to drive a turbine/generator to produce electricity and hot water, which can be used for direct heat applications. The steam, after passing through the turbine, is condensed in a heat exchanger and the excess heat is dissipated to the atmosphere through a cooling tower. The condensed geothermal fluid is returned to the deposit through an injection well to replenish the fluid reservoir.

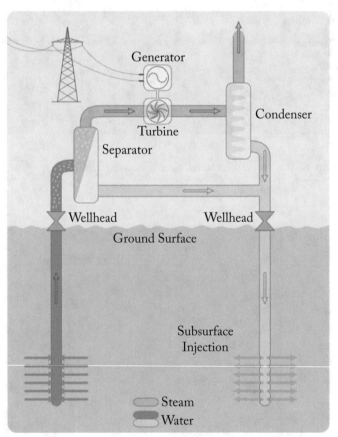

Figure 3.31: Diagram of a flash steam geothermal power plant. Based on Goran tek-en/Wikimedia Commons/CC-BY-SA-4.0, https://creativecommons.org/licenses/by-sa/4.0/deed.en, https://upload. wikimedia.org/wikipedia/commons/c/ce/Diagram_HotWaterGeothermal_inturperated_version.svg.

Figures 3.32–3.34 show images of geothermal generating stations. Figure 3.32 shows the borehole (production well) at the Krafla Geothermal Power Plant in Iceland. The borehole is located underneath the dome and the insulated pipes carry the geothermal fluid to the generating station. Figure 3.33 shows steam separators at the Hellisheiði Geothermal Power Station. Finally, Figure 3.34 shows the turbine building at the Reykjanes Geothermal Power Plant. The condensation associated with the air and water vapor exhaust from the cooling towers can be seen.

Figure 3.32: Borehole (under the brown dome) and insulated high pressure steam pipes at the Krafla Geothermal Power Plant in Iceland. Image courtesy of Richard A. Dunlap.

Figure 3.33: Steam separators at the Hellisheiði Geothermal Power Station in Hveragerði, Iceland. Image courtesy of Richard A. Dunlap.

Figure 3.34: Reykjanes Geothermal Power Plant in Gunnuhver, Iceland. Image: Jose Arcos Aguilar/ Shutterstock.com, https://www.shutterstock.com/image-photo/geothermal-field-gunnuhver-reykjanes-power-plant-236232832.

A number of countries have been active in developing geothermal energy for electricity production. Table 3.7 summarizes those countries with more than 200 MW installed geothermal electricity capacity as of 2015. While the U.S. has the greatest installed capacity, this represents only a small proportion of total electricity generation in that country. Some smaller nations, such as Kenya, Iceland, Philippines, and El Salvador, have been very serious in utilizing geothermal resources to fulfil a major component of their electricity needs. Figure 3.35 gives a breakdown of geothermal electricity worldwide by region and shows the development of this resource since 2005. Current geothermal electricity generation worldwide is around 75,000 GWh per year. Table 3.8 summarizes the use of direct use of geothermal energy worldwide. Major uses include space heating, bathing, aquaculture, and greenhouse heating.

Table 3.7: Leading nations in installed capacity of geothermal electricity generation

Country	Installed Geothermal Capacity (MW)	% Electricity Generation
United States	3450	0.3
Philippines	1870	27.0
Indonesia	1340	3.7
Mexico	1017	3.0
New Zealand	1005	14.5
Italy	916	1.5
Iceland	665	30.0
Kenya	594	51.0
Japan	519	0.1
Turkey	397	0.3
Costa Rica	207	14.0
El Salvador	204	25.0
Rest of world	656	-
Total	12,636	~0.3

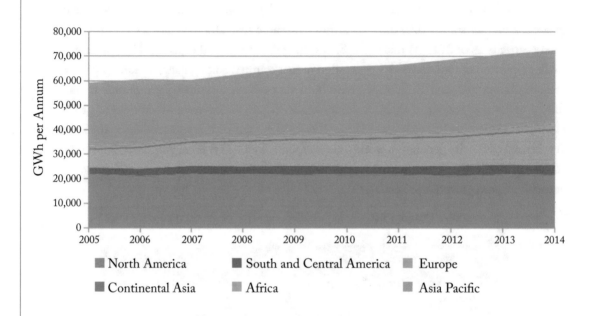

Figure 3.35: Growth of geothermal electricity in different regions of the world since 2005. Based on Figure 8 from The World Energy Council, London, UK, https://www.worldenergy.org/wp-content/uploads/2017/03/WEResources_Geothermal_2016.pdf.

Table 3.8: Installed capacity for direct use of geothermal energy by nation for 2014 for countries with more than 2 GW capacity. Data are from https://www.worldenergy.org/wp-content/uploads/2017/03/WEResources_Geothermal_2016.pdf

Country	Installed Capacity (MW)
China	17,870
United States	17,416
Sweden	5600
Turkey	2886
Germany	2849
France	2347
Japan	2186
Iceland	2040
Rest of world	16,823
Total	70,017

3.8   BIOFUELS

Biofuels may be solid, liquid, or gaseous and include a number of different energy sources ranging from wood, which is the oldest source of fuel used by humans, to biogas, which is produced from organic waste material. As seen in Figure 1.3, biofuels currently constitute nearly half of the renewable energy use in the U.S. Here the major categories of biofuels, which are likely to contribute to our sources of renewable energy, are overviewed.

3.8.1   WOOD

Wood has been used as a source of heat since antiquity and was the predominant source of energy for society worldwide until the late 19th century, when it was replaced by coal. Today, wood is used primarily in industry for manufacturing paper products and building construction. However, as Figure 1.3 shows, it also contributes to renewable energy sources in the U.S. In a number of less industrialized regions of the world, i.e., Bangladesh, wood is still the major fuel used for heating and cooking.

Carbon in wood produces carbon dioxide when it is burned by the reaction

$$C + O_2 \rightarrow CO_2 . \tag{3.14}$$

Trees sequester carbon dioxide from the atmosphere by photosynthesis, where CO_2 is combined with water to yield complex hydrocarbons. In principle, the use of wood as a source of energy is carbon neutral and renewable, if trees are replanted as they are consumed. This is generally the case in North America but is often not the case in other parts of the world.

Much of the wood that is used as fuel is used for residential space heating. However, in recent years wood has also been used for grid-scale electricity generation. A number of formerly coal-fired generating stations worldwide have been converted to operate on wood pellets. One such example is shown in Figure 3.36. This is the Atikokan Generating Station in Ontario, Canada. The station was constructed as a coal-fired generating station in 1985 and was converted to wood-pellet use in 2014. With a nameplate capacity of 205 MW capacity it is largest exclusively wood-fired generating station in North America. The two cylindrical structures to the right of the station in the photograph, are silos, 21 m in diameter and 44 m tall, with a capacity of 5,000 tons of wood pellets.

Figure 3.36: Atikokan Generating Station in northwestern Ontario, Canada (reproduced with permission). https://www.opg.com/generating-power/thermal/stations/atikokan-station/Pages/atikokan-station.aspx.

3.8.2   MUNICIPAL SOLID WASTE

Municipal solid waste (MSW) is defined as non-hazardous solid human-generated waste material. It includes durable materials, such as glass and metals, as well as nondurable materials, such as paper and food waste. While the former component of MSW can often be recycled, the latter component is, potentially, useful for generating energy. Per capita MSW generation varies enormously (i.e., by a factor of 50 or more) between countries, as illustrated by the examples in Table 3.9. There is some correlation between per capita waste generation and per capita gross domestic product, with more industrialized countries typically generating around 2 kg per person per day of MSW. The average energy content of MSW is in the vicinity of 10 MJ/kg, meaning that the per capita MSW energy is typically around (10 MJ/kg) × (2 kg) = 20 MJ per day. This amounts to an average power of

$$(20 \text{ MJ})/(24 \text{ h} \times 3600 \text{ s/h}) = 230 \text{ W} . \tag{3.15}$$

Recall from Chapter 1 that the average per capita power consumption in North America is about 12 kW. So, the energy content from MSW accounts for about (12 kW)/(230 W) = 0.019, or about 2% of our energy consumption.

Table 3.9: Municipal solid waste in kg per capita per day for some selected countries (data are from http://sitere-sources.worldbank.org/INTURBANDEVELOPMENT/Re-sources/336387-1334852610766/AnnexJ.pdf)

Country	MSW (kg/capita/day)
Antigua	5.50
Sri Lanka	5.10
Ireland	3.58
Norway	2.80
United States	2.58
Canada	2.33
France	1.92
United Kingdom	1.79
Japan	1.71
Iceland	1.56
China	1.02
Russia	0.93
India	0.34
Uruguay	0.11

Figure 3.37 gives a breakdown of disposal methods of MSW in the U.S. As the figure shows, 12.8% of MSW produced in 2015 was used for energy recovery by combustion. Two approaches can be used for energy recovery from MSW: direct combustion and combustion of methane gas produced at landfill sites by the decomposition of MSW.

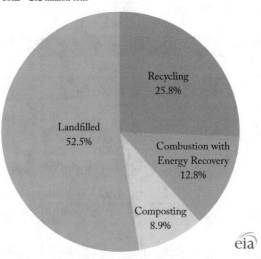

Management of MSW in the United States, 2015

Total = 262 million tons

Source: U.S. Environmental Protection Agency.
Advancing Sustainable Materials Management: 2015 Fact Sheet, July 2018

Figure 3.37: Breakdown of MSW disposal methods in the U.S. in 2015. Based on https://www.eia.gov/ energyexplained/?page=biomass_waste_to_energy.

3.8.3   ETHANOL

Ethanol (C_2H_5OH) is a particularly convenient biofuel as it is a liquid at room temperature and can be used as a direct replacement for gasoline in internal combustion engines with very minimal modifications. The production of ethanol begins with the conversion of carbon dioxide and water to glucose ($C_6H_{12}O_6$) by photosynthesis:

$$6CO_2 + 6H_2O \rightarrow C_6H_{12}O_6 + 6O_2 \; . \tag{3.16}$$

This reaction is endothermic and occurs because of the input of energy from sunlight. The production of ethanol follows from the fermentation of glucose by the reaction

$$C_6H_{12}O_6 \rightarrow 2C_2H_5OH + 2CO_2 \; . \tag{3.17}$$

Energy is extracted from ethanol by the combustion process

$$C_2H_5OH + 3O_2 \rightarrow 2CO_2 + 3H_2O \; . \tag{3.18}$$

An analysis of these three reactions shows that $6CO_2$ are used in the reaction in Equation (3.16) while $2CO_2$ are released in Equation (3.17) (to produce $2C_2H_5OH$) and $4CO_2$ are released in the combustion of $2C_2H_5OH$ according to Equation (3.18). The overall process is, therefore, carbon neutral.

Ethanol production has increased substantially since around 2000, as shown in Figure 3.38, with the greatest increase in ethanol used for energy production. The U.S. now leads in ethanol production, as shown in Figure 3.39, followed by Brazil, which has made a major effort to implement ethanol as a transportation fuel.

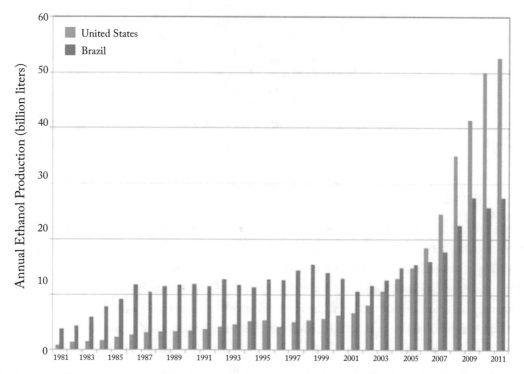

Figure 3.38: Annual ethanol production in the U.S. and Brazil from 1981–2011. Units are 10^6 liters. Based on Figure 1 from Wang et al., 2012.

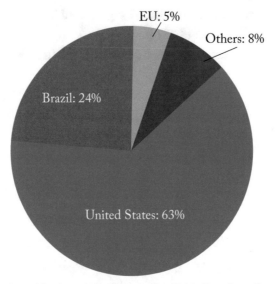

Figure 3.39: Breakdown of world ethanol production for 2011. Based on Renewable Energy Network, 2005–2012 Renewables Global Status Reports, Paris, France: REN21 Secretariat, Timilsina, 2013.

3.8.4   BIODIESEL

While ethanol is a more-or-less direct replacement for gasoline, biodiesel is a direct replacement for petroleum-derived diesel fuel. Biodiesel is manufactured by the transesterification of oils from vegetable or animal sources. The availability of suitable oils is a major problem for the production of quantities of biodiesel fuel suitable for replacement of petroleum-derived fuels. Waste or low-grade vegetable and animal oils from the food industry can supply only a small fraction of starting material to fulfil biodiesel needs, and recovery is often inconvenient and expensive. Some potential plants for biodiesel production are given in Table 3.10. It is clear that tropical plants, such as palm, are more productive than plants from temperate regions, such as rapeseed and sunflower. Algae, however, yields the highest production of biodiesel per farming area. Algae also has the advantage that it can be grown in shallow marine or pond environments and does not decrease the available arable land for food production.

Table 3.10: Biodiesel productivity in 1000 L/km^2 per year for some plant materials (data adapted from Dunlap, 2019)

Plant	$10^3\ L/km^2/y$
Sunflower	77
Rapeseed (canola)	95
Palm oil	475
algae	1700

Figure 3.40 shows that biodiesel production has increased since about 2007 but is still much less than annual ethanol production. Figure 3.41 shows that the European Union, is the most active in producing biodiesel fuel.

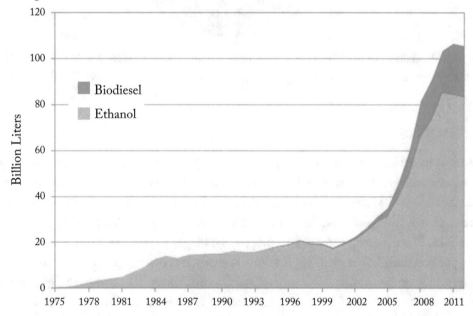

Figure 3.40: World biodiesel production compared with world ethanol production from 1975–2012. Based on F.O. Licht, REN21, http://www.worldwatch.org/biofuel-production-declines-1.

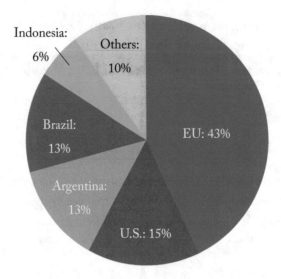

Figure 3.41: Breakdown of world biodiesel production for 2011. Based on Renewable Energy Network, 2005–2012 Renewables Global Status Reports, Paris, France: REN21 Secretariat, Timilsina, 2013.

3.8.5   BIOGAS

Biogas refers to a mixture of combustible gases which is produced by the anaerobic decomposition of organic matter. Feedstock used to produce biogas most frequently consists of cow manure and/or agricultural waste. Material is decomposed in a light-free and oxygen-free environment in a digester as shown in Figure 3.42. The typical composition of the resulting biogas is summarized in Table 3.11. It is generally necessary to process the biogas to increase the methane content by removing carbon dioxide. For many applications, however, it is most important to remove the small hydrogen sulfide component, as this is very corrosive to machinery. The resulting biogas can be used as a gaseous fuel to substitute for natural gas, which is mostly methane. Frequently, however, biogas production facilities use an internal combustion engine/generator to produce electricity.

Figure 3.42: Digesters at a biogas plant in Germany. Image: Lena Wurm/Shutterstock.com, https://www.shutterstock.com/image-photo/biogas-plant-germany-405953155?src=E-BEPcl-p2qlJX-sIRoPXEw-1-3.

Table 3.11: Typical composition of biogas produced in a digester by anaerobic decomposition (data adapted from https://web.archive.org/web/20100106022729/ http://www.kolumbus.fi/suomen.biokaasukeskus/en/enperus.html)

Gas	Formula	Typical Composition (%)
Methane	CH^4	55–75
Carbon dioxide	CO_2	25–45
Nitrogen	N_2	1–5
Hydrogen	H_2	0–3
Hydrogen sulfide	H_2S	0.1–0.5
Carbon monoxide	CO	0–0.3

3.9   SUMMARY

The assessment of the suitability of renewable energy resources as described above for the long-term replacement of fossil fuels requires a consideration of several factors as outlined earlier in the present chapter. Here we summarize renewable energy sources on the basis of three important criteria: cost, environmental impact, and human risk.

3.9.1   COST

Figure 3.43 shows the estimated cost for electricity in the U.S. produced by different methods. Conventional coal, which averages $95/MWh, is a relevant basis for comparison, as this primary energy source has traditionally provided a substantial fraction of the base load electricity in the U.S.

Figure 3.43 shows that electricity generated from natural gas by a combined cycle power plant is somewhat less expensive than traditional coal. The combined cycle system utilizes two turbines operating in tandem. The first is a combustion turbine utilizing natural gas as a fuel. The waste heat from the combustion turbine is used to produce steam to drive the second turbine. This type of tandem system can yield efficiencies of up to 62% compared to the efficiency of a single stage heat engine of 35–40%. The use of natural gas in a combustion turbine, as indicated in the figure, is less economical than the combined cycle system. Combustion turbines are in common use as a means of topping-up base load in times of increased demand because they can be brought on-line much more quickly than systems that rely on thermal generation (i.e., coal, nuclear, combined cycle, etc.). The figure also shows that the cost for electricity generated by nuclear facilities is virtually the same as for coal-fired stations.

Electricity from renewable sources covers a very large range of costs, nearly a factor of 10 from the low end of geothermal to the high end of solar thermal. Geothermal, onshore wind, and hydroelectric generation is generally more economical than traditional coal, while biofuels are about the same. Solar photovoltaics are, on average, slightly more expensive than coal, but there is a considerable range of prices associated with this technology. It may be hoped that further research and development of photovoltaic devices will stabilize the cost and bring it down to a level consistent with other energy sources. Offshore wind is, for the foreseeable future, an expensive option because of infrastructure and maintenance costs associated with the offshore location. Solar thermal, as discussed above, is in the relatively early stages of technical development and consists of several distinctly different technologies.

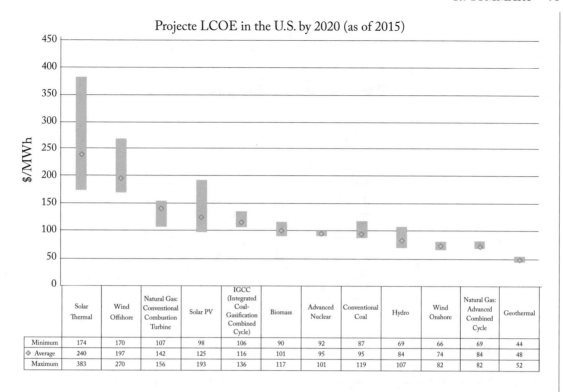

Figure 3.43: Projected cost of electricity produced by various methods in the U.S. in 2020. Data are from https://www.eia.gov/outlooks/aeo/. https://commons.wikimedia.org/wiki/File:Projected_LCOE_in_the_U.S._by_2020_(as_of_2015).png. Based on Andynct/Wikimedia Commons/CC-BY-SA 4.0, https://creativecommons.org/licenses/by-sa/4.0/deed.en.

3.9.2   CARBON FOOTPRINT

All methods of producing energy will have some environmental impact, including the emission of greenhouse gases, notably CO_2. In some cases, these emissions result from manufacturing processes associated with the construction of equipment, while in other cases environmental changes are responsible for net greenhouse gas emission. Numerous studies have quantified CO_2 emissions from various energy sources. Results of this type of analysis are summarized in Table 3.12. For fossil fuels, such an analysis is fairly straightforward, as the energy which is generated and the CO_2 emissions are both directly related to the quantity of fuel consumed. For renewable energy sources, the estimated CO_2 emissions are based on a consideration of emissions from infrastructure manufacturing and maintenance (such as mining and concrete production) and emissions from environmental consequences (such as deforestation).

Table 3.12: Greenhouse gas (CO_2) emissions for different methods of generating electricity (data adapted from Nugent and Sovacool 2014)

Method	kg(CO_2)/MWh
Coal	1050
Oil	778
Natural gas	443
Nuclear	66
Photovoltaic	50
Geothermal	41
Wind	34
Solar thermal	13
Biogas	11
Hydroelectric	10

3.9.3   RISK

The risk to humans that results from various methods of producing energy is difficult to quantify. The estimated mortality rate for electricity generated by different techniques as measured by deaths per unit of electrical energy is often used as a means of comparison. Although this type of assessment is not straightforward, two sources of difficulty are apparent in this type of analysis. First, total mortality for some electricity generating methods is fairly obvious, for example for wind energy as discussed above, as deaths are the immediate result of accidents. In other cases, for example coal, where the majority of deaths may result for long-term exposure to pollutants, it is much more difficult to correlate deaths and specific energy sources. Second, mortality for some energy sources is dominated by a small number of events, such as the 1975 hydroelectric dam failures in China or the 1986 nuclear accident at the Chernobyl reactor. It is difficult to assess whether these incidents should be considered anomalies or whether one might expect similar events to occur periodically.

Table 3.13 summarizes human mortality for different energy sources and, in some cases, for different regions. For coal, the mortality rate worldwide is dominated by coal-fired generation in China. The rate for the U.S. is much lower as a result of more stringent pollution regulations. The risk associated with biofuels is surprisingly high. This situation is dominated by traditional wood consumption in less developed countries and is not representative of advanced biofuels such as ethanol, biodiesel and biogas. The worldwide risk associated with hydroelectric and nuclear energy is dominated by the events mentioned above. The risk associated with these energy sources, as based on historical events in the U.S., is substantially lower.

Table 3.13: Estimated human mortality rate in deaths per PWh (1015 Wh) of electricity generated for different energy sources and different regions as described in the text. Estimates are from 2012 (data adapted from https://www.forbes.com/sites/jamesconca/2012/06/10/energys-deathprint-a-price-always-paid/#2767b73e709b)

Energy Source	Region	Deaths per PWh
Coal	World	100,000
Coal	U.S.	10,000
Biofuels	World	24,000
Natural gas	World	4000
Solar	World	440
Wind	World	150
Hydroelectric	World	1400
Hydroelectric	U.S.	5
Nuclear	World	90
Nuclear	U.S.	0.1

The above evaluations of cost, carbon emissions, and risk provide an interesting comparison of different renewable energy sources and fossil fuels. Geothermal, onshore wind, hydroelectric, and (most likely) solar photovoltaics can (or should in the near future) compete fairly favorably with traditional fossil fuels. A carbon footprint analysis shows that renewable energy sources typically contribute less than about 10% of the greenhouse gases of fossil fuels. In addition, a risk assessment shows that the human mortality rate associated with renewable energy (and also nuclear energy) is substantially less than for fossil fuels. This is true, even when individual (and perhaps anomalous) disasters are included in the analysis.

CHAPTER 4

The Need for Energy Storage

4.1 INTRODUCTION

Chapter 2 presented the need for our energy use to move away from fossil fuels and for the implementation of carbon-free renewable energy sources. Chapter 3 presented the major approaches to renewable energy production that are likely to play an important role in our energy use in the relatively near future. The renewable energy sources discussed in Chapter 3 fall into basically three categories. In most cases, the energy which is produced is electrical energy. In the case of the direct use of solar or geothermal energy, the energy is in the form of heat. Finally, biofuels generally produce a fuel where the energy is stored as chemical energy in the bonds between atoms in the material.

Our energy use is not uniform in time or space. Fossil fuels have generally been a convenient means of dealing with this situation. Electricity, generated by the combustion of coal and natural gas, is, in most cases, able to provide power on-demand as needed. Gasoline and diesel fuel have a high energy density and are a very convenient portable energy source for transportation and for use in remote areas. However, renewable energy that is produced by the methods discussed in the previous chapter is not necessarily available when and where it is needed and often requires a storage mechanism to allow it to be utilized effectively. The present chapter reviews the specific needs for energy storage and overviews the availability of energy from the major renewable sources.

4.2 DISTRIBUTION OF ENERGY USE IN TIME

4.2.1 ELECTRICITY

Monitoring of grid-based electricity use is the simplest means of assessing real-time energy use. Figure 4.1 shows typical energy use throughout the week. Daily electricity use is similar during the week (Monday to Friday). Electricity use is less on the weekend due to reduced business/commercial/industrial use.

Seasonal variations can be seen in Figure 4.2 where fluctuations in electricity use in different sectors are illustrated. The greatest seasonal variations occur in the residential sector, where peak use in the summer results from air conditioning and peak use in the winter comes from heating and additional lighting.

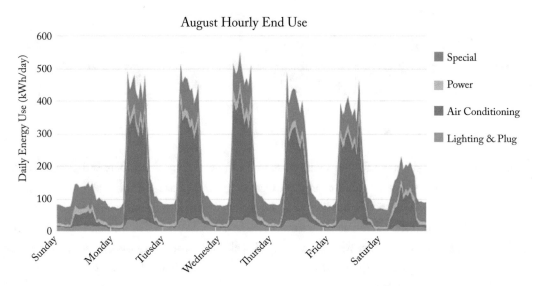

Figure 4.1: Weekly electricity use at the Shenzhen Institute of Building Research (IBR). Based on Diamond, et al., 2013.

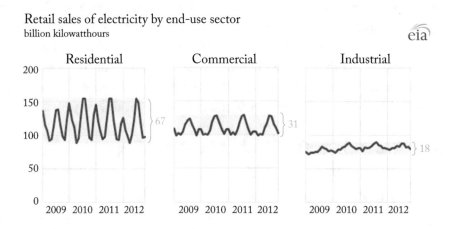

Figure 4.2: Seasonal variations in electricity use in the U.S. for residential, commercial, and industrial sectors. Based on https://www.eia.gov/todayinenergy/detail.php?id=10211, U.S. Energy Information Administration.

In the summer (i.e., July in the Northern Hemisphere and January in the Southern Hemisphere), daily electricity use shows a single peak, as illustrated in Figure 4.3, with a maximum around the middle of the afternoon. Air conditioning plays a significant role in this energy use. In the winter (i.e., January in the Northern Hemisphere and July in the Southern Hemisphere), the figure illustrates that daily electricity use shows two peaks, one in the morning and one in the early evening, due to additional lighting and heating during those periods. This figure shows the break-down of electricity use into base load (which meets the minimum requirements at all times during the day), intermediate load and peak load (for the highest usage times). The use of energy storage during low demand periods to supplement energy production during high demand periods helps to reduce the need for additional electricity generating capacity. Such an approach may be referred to as "load leveling," "peak shaving" or "peak smoothing."

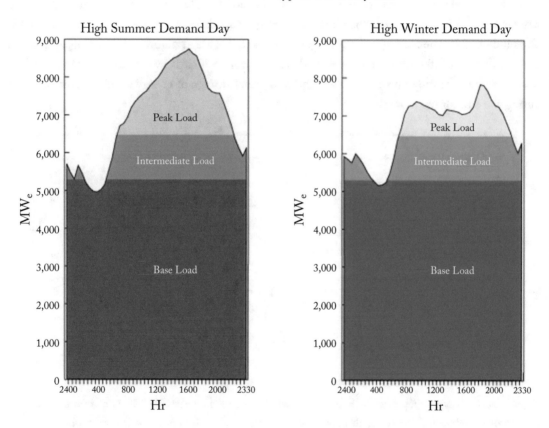

Figure 4.3: Typical workday electricity consumption during the summer (left) and during the winter (right). Based on http://www.world-nuclear.org/information-library/energy-and-the-environment/renewable-energy-and-electricity.aspx.

4.2.2   THERMAL ENERGY

Although electricity generated by renewable means can be used for heating (electric heat), ther-mal energy produced by some renewable sources (e.g., solar thermal and geothermal) can be used directly for space heating purposes. It is, therefore, of relevance to look at heating requirements as a function of time. As with electrical requirements (some of which contribute to heating, as well), daily, as well as seasonal, variations should be considered.

Heating requirements, of course, depend on local climate, where requirements are largest in cold regions and smallest in tropical regions. Cooling requirements (i.e., air conditioning), which are satisfied using electricity, are just the opposite. In many parts of the world both heating and cooling are required seasonally.

Figure 4.4 shows temperature in a temperate region (i.e., Denmark) as a function of time of day for a period in May. Heating is required when the outside temperature drops below about 18° C, as indicated by the red regions on the graph. These times are at night, between roughly 18:00 h and 08:00 h. Thus, the utilization of solar thermal energy, which is available during the day (see Section 4.3) for heating purposes would require thermal storage of energy acquired from sunlight during the day, for use at night.

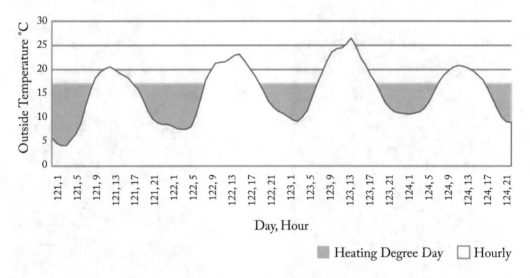

Figure 4.4: Hourly temperature for four days in May (days of the year 121–124, with hours indicated) in Gentofte, Denmark. The graph also shows time when heat is required (in red) for a typical building. Based on Cox, et al. (2015), reprinted with permission from Elsevier.

It is also of interest to consider the seasonal variations in heating requirements. Figure 4.5 shows heating and cooling requirements for a building in Denmark, throughout the year. As expected, heating requirements are greatest during the winter when the availability of solar thermal energy is at a minimum (see Section 4.3).

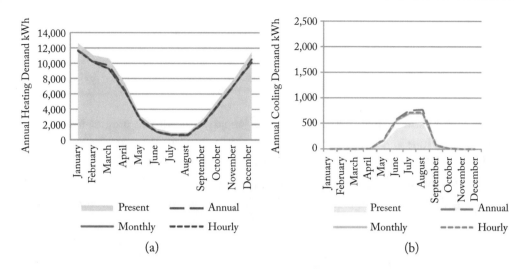

Figure 4.5: (a) Heating requirements and (b) cooling requirements as a function of month for a building in Gentofte, Denmark. Based on Cox et al. (2015), reprinted with permission from Elsevier.

4.3   TEMPORAL VARIATIONS IN RENEWABLE ENERGY

It is clear from the previous section that our energy use, both electricity and thermal energy, is not uniform in time. If renewable energy satisfies a major component of our energy use, it is important to see how the availability of different sources of renewable energy varies in time. As overviewed in the previous chapter, renewable sources may produce energy in one of three forms:

1. electricity directly (e.g., hydroelectricity, wind, photovoltaics),

2. heat (solar thermal, geothermal), and

3. fuel with stored chemical energy (biofuels).

In general, biofuels can be used in much the same way as solid, liquid, or gaseous fossil fuels. This includes thermal generation of electricity (e.g., to replace coal) and liquid or gaseous fuel for internal combustion engines. Typically, such systems provide energy on-demand and the ability to store energy is not a major concern. Therefore, in the present section we consider the properties of the other major renewable energy sources: hydroelectric, wind, solar, and geothermal.

4.3.1   HYDROELECTRIC ENERGY

Although hydroelectric generation is fairly consistent on a day to day basis, there are seasonal fluctuations that result from fluctuations in river flow rates. These variations are readily seen in Figure 4.6, which shows the total monthly hydroelectric generation in the U.S. over a period of five years.

Data consistently show a small maximum around January, a minimum around March, a major maximum around June and a second minimum around September. These are the result of annual climatic variations that are reasonably consistent from year to year and can account for differences of a factor of two or so in hydroelectric output during the year. Figure 4.7 shows flow rates of two Austrian rivers that illustrate similar features. The data clearly show annual maxima around June of each year.

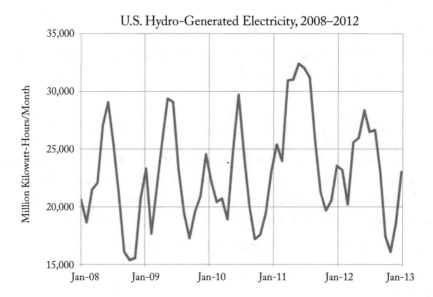

Figure 4.6: Monthly averaged hydroelectric energy generation in the U.S. for 2008–2012. Data from http://www.eia.gov/electricity/data.cfm#generation, based on Plazak/Wikimedicommons/CC-BY-SA3.0, https://creativecommons.org/licenses/by-sa/3.0/deed.en). https://commons.wikimedia.org/wiki/File:US_Monthly_Hydro_Power_Generation.png.

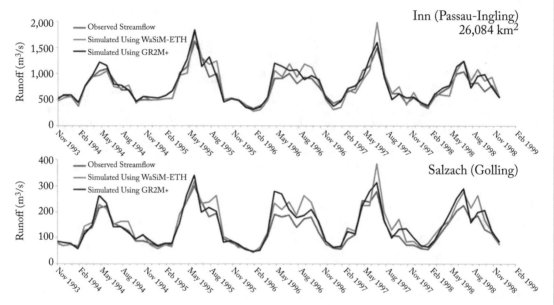

Figure 4.7: Monthly averages of flow rates in two Austrian rivers (Inn River, top, and Salzach River, bottom) over a period of about five years. Passau-Ingling and Golling are the locations of hydroelectric facilities on the Inn River and Salzach River, respectively. The red lines show measured flow rates and the blue and black lines show flow rates as calculated by different models. Based on figure 4 from Wagner et al. (2017).

4.3.2   WIND ENERGY

Figure 4.8 shows some wind power measurements made over a two-week period for wind turbines in South Australia. While there are clear variations in wind intensity over this period, there are no clear systematic variations on a daily or weekly time scale. However, averaged over a longer time period, there are more predictable variations in wind intensity. This is illustrated in the data shown in Figure 4.9 for normalized wind energy at the Port Burwell Wind Farm in Ontario, Canada, where wind energy shows a minimum during the summer months. The same seasonal systematics are illustrated for wind generated electricity in the U.S., as shown in Figure 4.10.

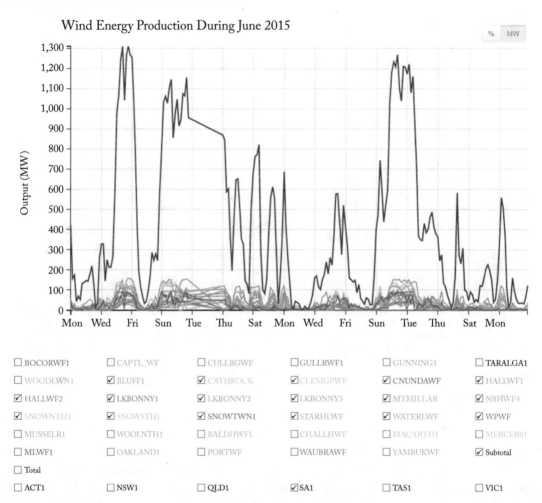

Figure 4.8: Wind power at a number of turbines in South Australia over a two-week period. The total power for all turbines is shown by the black line. Based on https://stopthesethings.files.wordpress.com/2015/07/june-2015-sa.png; reproduced with permission.

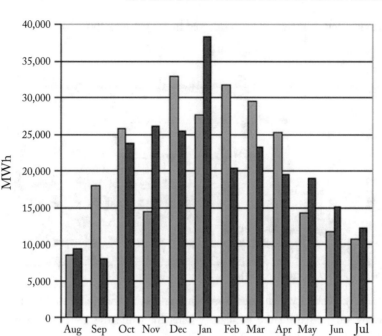

Figure 4.9: Monthly wind energy during two years at the Port Burwell Wind Farm (formerly known as the Erie Shores Wind Farm) in Ontario, Canada. Based on https://commons.wikimedia.org/wiki/File:Erie_Shores_Wind_Farm_output_Aug-Jul_2008.gif.

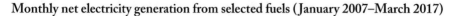

Monthly net electricity generation from selected fuels (January 2007–March 2017)

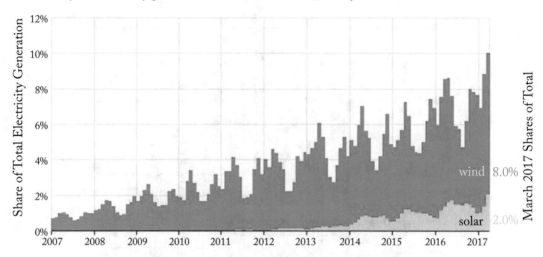

Figure 4.10: Monthly electricity generated from wind and solar energy in the U.S. from 2007 to 2017. Seasonal variations are seen from the oscillatory behavior while the general increasing trend results from the overall increase in generating capacity. Based on https://www.eia.gov/todayinenergy/detail. php?id=31632, U.S. Energy Information Administrtion.

4.3.3   SOLAR ENERGY

Daily as well as seasonal variations in the solar energy arriving at the surface of the earth are obvious. It is also obvious that non-periodic variations resulting from local weather conditions also exist. Figure 4.11 shows the daily variations in power output of 1 kW of photovoltaic panels in Los Angeles averaged over the year. The figure also shows the effects of collector orientation and tracking. The seasonal fluctuation in average daily solar energy is illustrated in Figure 4.12. It is seen in these graphs that, in addition to the variations in the insolation at the surface of the earth that occurs throughout the day, there is a difference of about a factor of four between summer and winter at 43° north latitude. Seasonal variations are greater at higher latitudes and less at lower latitudes.

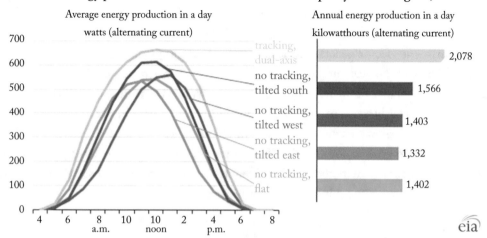

Figure 4.11: Average daily variations in the output of 1 kW of photovoltaic panels in Los Angeles, CA (34° north latitude) for different collector orientations. Based on https://www.eia.gov/todayinenergy/detail.php?id=18871, U.S. Energy Information Administrtion.

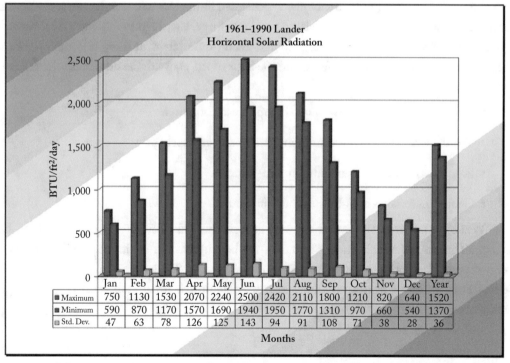

Figure 4.12: Maximum and minimum daily solar energy as a function of month for the period of 1961–1990 at Lander, Wyoming (~43° north latitude). Based on and reproduced with permission from J. Curtis and Kate Grimes, http://www.wrds.uwyo.edu/sco/climateatlas/solar.html.

4.3.4   GEOTHERMAL ENERGY

While one might not expect variations in electrical output from geothermal generating facilities, this is not entirely the case. Electricity is produced from geothermal resources by means of a heat engine that drives an electric generator. The efficiency of such a system is governed by the thermodynamic efficiency of the heat engine, which is most easily approximated by the Carnot efficiency based on the temperatures of the hot and cold reservoirs. While the temperature of the geothermal deposit may remain fairly constant, the temperature of the cold reservoir, which is most commonly the atmosphere, will show systematic seasonal variations. These variations (between winter and summer months) are greatest for facilities that utilize relatively low temperature geothermal resources and can amount to more than a 20% difference in efficiency (and subsequently, in capacity factor) between maximum and minimum output periods.

4.3.5   SUMMARY

Figure 4.13 shows the monthly electricity generation in the U.S. from non-fossil fuel sources. Based on current proportions of electricity generated by renewable sources in the U.S., monthly and seasonal fluctuations are dominated by the variability of hydroelectric generation. The proportions of renewable energy (excluding biofuels) generated in the U.S. and (on the average) elsewhere worldwide are weighted toward hydroelectric and wind (see, e.g., Figures 1.3 and 1.11). However, in general, the different renewable energy sources tend (some extent, at least) to complement each other in their seasonal variations. We can, on the basis of the above discussion, make the following observations.

- Hydroelectric energy shows a maximum in early summer.

- Wind energy shows a maximum in winter and early spring.

- Solar energy shows a maximum in mid-summer.

- Geothermal shows a slight maximum in winter.

While some fraction of hydroelectricity and geothermal electricity may contribute to base load requirements, renewable energy resources are best utilized in conjunction with a means of energy storage, in order to best match fluctuations in demand. Similarly, thermal energy from solar thermal installations or from geothermal resources requires thermal storage to best meet space heating needs.

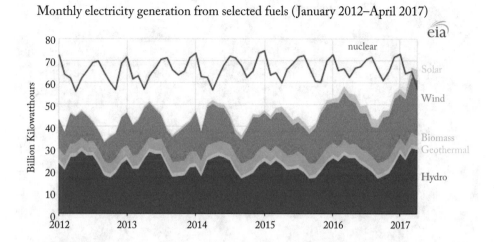

Figure 4.13: Monthly electricity generation in the U.S. from different sources between January 2012 and April 2017. Based on https://www.eia.gov/todayinenergy/detail.php?id=31932., U.S. Energy Information Administration, *Monthly Energy Review and Electric Power Monthly*.

4.4   REMOTE AND EMERGENCY ENERGY SYSTEMS

Facilities that require electricity but cannot be connected to the electric grid because of location, or those that require electricity when there is a grid failure, need alternative methods of generating and/or storing electricity. In the past, fossil-fuel powered generators (usually diesel), as shown in Figure 4.14, were generally used, but future trends away from energy sources that contribute to greenhouse gas emissions will require the implementation of renewable energy sources and will likely need to incorporate suitable energy storage.

Figure 4.14: Emergency diesel generator used to provide backup electricity for a research facility. Image courtesy of Richard A. Dunlap.

Small-scale remote energy systems include facilities such as remote communications antennas as shown in Figure 4.15. Here solar and wind resources are used in conjunction with battery storage to provide continuously available power.

Figure 4.15: Remote radio antenna in Iceland using a combination of a wind turbine (to the left of the antenna) and solar photovoltaics (panels mounted on the antenna) along with battery storage (boxes at the base of antenna). Image courtesy of Richard A. Dunlap.

Mid-sized systems are often in the 100 kW range and include typical emergency power systems for applications such as hospitals, telecommunication centers, and police stations. Alternatives to traditional fossil fuel powered generators such as wind or solar systems require appropriate energy storage methods. Systems may also utilize non-fossil fuel chemical energy storage methods such as hydrogen or ammonia to produce energy through fuel cells or internal combustion generators. Figure 4.16 shows a typical system utilizing hydrogen fuel cells.

Figure 4.16: Emergency energy system using hydrogen-powered fuel cells for a telecommunications facility, https://commons.wikimedia.org/wiki/File:FuelCellSystem.jpg.

Large-scale facilities in the MW range can be used to provide power to remote villages or islands. As shown in Figure 4.17, renewable energy sources such as wind and solar, including appropriate storage methods, may be used.

Figure 4.17: Remote power station on King Island (population ~2000) off the coast of Tasmania. Wind, solar photovoltaics, and biodiesel generators provide renewable energy in conjunction with traditional diesel backup generators and battery storage. Photo courtesy Hydro Tasmania, http://www.kingisland-renewableenergy.com.au/stand-alone-power-systems/what-hybrid-grid-power-system.

4.5 PORTABLE ENERGY SYSTEMS

Typical small-scale portable applications requiring energy storage include personal electronics, such as calculators, cell phones and computers. These generally use batteries as an energy storage medium, occasionally with photovoltaic cells to supplement or charge the batteries from ambient light.

Portable energy systems, such as those needed for vehicles are, perhaps, the greatest challenge for the implementation of carbon-free energy. Energy sources for vehicles need to have sufficient energy density and power density that they are small enough and light enough to be contained within a practical vehicle. Gasoline and diesel fuel have been the choice for vehicle fuels for more than the past century. Their high energy density, about 50 MJ per kg, means that a reasonable fuel mass (say 50 kg for a passenger vehicle) will provide a vehicle range of 600–800 km or more, and their combustion in a typical internal combustion engine will yield sufficient power for most vehicle applications.

Biofuels can provide a more-or-less direct replacement for liquid fossil fuels. Table 4.1 gives some properties of gasoline and diesel fuel, along with those of biofuel alternatives. As can be seen in the table, the energy content, both in terms of fuel mass and in terms of fuel volume, are less for biofuels than for the corresponding fossil fuels. This is particularly notable in the case of ethanol compared with gasoline. This means that the fuel requirements to achieve the same vehicle range will be somewhat greater. Similarly, the fuel consumption (sometimes measured as L/100 km) will be greater for vehicles utilizing gasoline-ethanol mixtures (sometimes referred to as "flexfuel"), than for the same vehicle using pure gasoline.

Table 4.1: Density and energy content of gasoline, diesel fuel and ethanol. Note values for gasoline, diesel, and biodiesel are approximate, as the exact composition of these fuels varies

Fuel	Density (kg/m^3)	Energy (MJ/kg)	Energy (MJ/L)
Gasoline	735	45.8	33.7
Diesel	811	45.5	36.9
Ethanol	789	26.8	21.2
Biodiesel	912	37.8	34.5

Other than biofuels, renewable energy sources are not suitable as a direct means of powering a practical vehicle, either because of lack of portability or low energy density, or both. While experimental solar powered vehicles, as shown in Figure 4.18, have been constructed, they are not practical for conventional road use. At noon on a sunny day (at 45° latitude), the horizontal insolation is around 600 W/m^2. A typical experimental solar vehicle, as in the figure, would have an active photovoltaic area of about 5 m^2. For a photovoltaic efficiency of 20% (typical of commercial silicon cells) the net power available would be about 600 W (or in traditional units, a bit less than 1 hp). Thus, a portable method of storing the electricity produced by renewable energy sources is necessary to power a practical road vehicle.

A convenient means of presenting information about energy storage methods is the Ragone plot (pronounced ru-GO-nee) as shown in Figure 4.19. This plot is named after the American metallurgist David V. Ragone (b. 1930) who is best known for the plot, which bears his name, and for his position as President of Case Western University (1980–1987). The Ragone plot is a graph of specific power (sometimes called power density) as a function of specific energy (sometimes called energy density). This graph is sometimes drawn with the horizontal and vertical axes interchanged. In terms of vehicle applications, high specific energy is desirable, as it provides greater range for the same mass of energy carrier. High specific power is also desirable as it provides maximum available power for a given energy carrier mass.

Figure 4.18: Solar-powered vehicle "Emilia 2" at Expo' AEM-ZERO Alternative Energy for Mobility Zero Emission 2013 in Lugo, Italy. Image: Shutterstock.com, https://www.shutterstock.com/image-photo/lugo-ra-italy-september-19-unidentified-167237564?src=8_H9TMWI_uulRzHA4-hUPg-1-3.

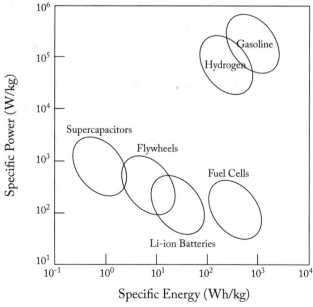

Figure 4.19: Ragone plot for some energy carriers that may contribute to energy requirements for transportation.

It is seen in the Ragone plot in Figure 4.19 that most of the energy storage technologies fall along a line with negative slope near the bottom of the graph. This means that carriers that have a large energy storage capacity provide less power, that is, the energy cannot be extracted quickly. On the other hand, carriers that release energy quickly, and therefore, supply large amounts of power, tend to have lower storage capacity. Unfortunately, for vehicle applications, this means that there is a trade-off between available power and available range.

As shown in the plot, gasoline and hydrogen are separated from this line and appear in the upper part of the figure. Based on the discussion above concerning biofuels, ethanol would fall in the region between gasoline and hydrogen on the graph. The much lower specific energy for most forms of renewable energy storage compared with gasoline is a major challenge (although not necessarily the only one) for implementing carbon-free transportation technologies.

4.6   SUMMARY

The present chapter has shown that energy from renewable sources is not always available when or where it is needed. The storage of renewable electrical or thermal energy is necessary to effectively satisfy our needs such as grid electricity, residential space heating, remote off-grid systems, and transportation. For grid electricity, the daily and/or seasonal fluctuations in renewable energy need to be accommodated in order to satisfy the variations in our electrical needs. The same is true for residential heating requirements. At present, remote locations and facilities that require emergency electrical backup power most commonly use fossil fuel-based generating systems. The transition to renewable sources requires the integration of energy storage methods to provide continuous and reliable electricity. Portable energy requirements, particularly as needed for transportation, are a challenging aspect of energy storage because of the low-energy and/or low-power density of many storage methods. Overall, the suitability of specific energy storage methods must be considered on the basis of total energy capacity and available power, as well as cost.

Bibliography

Aubrecht, G. J. (2006). *Energy: Physical, Environmental and Social Impact*, 3rd ed. Upper Saddle River, NJ: Pearson Prentice Hall. 2

Bloom, A. J. (2010). *Global Climate Change: Convergence of Disciplines*. Sunderland, MA: Sinauer Associates.

Boyle, G., Ed., (2012). *Renewable Energy: Power for a Sustainable Future*, 3rd ed. Oxford: Oxford University Press.

Brecha, R. J. (2012). "Ten reasons to take peak oil seriously," *Sustainability* 5, pp. 664–694. DOI: 10.3390/su5020664. 21

Canada's Energy Future (2016). Energy Supply and Demand Projections to 2040, https://www.neb-one.gc.ca/nrg/ntgrtd/ftr/2016/index-eng.html.

Cavallo, A. J. (2004). "Hubbert's petroleum production model: An evaluation and implications for world oil production forecasts," *Natural Resources Research* 13, pp. 211–221. DOI: 10.1007/s11053-004-0129-2.

Cox, R. A., Drews, M., Rode, C., and Nielsen, S. B. (2015). "Simple future weather files for estimating heating and cooling demand," *Building and Environment* 83, pp. 104–114. DOI: 10.1016/j.buildenv.2014.04.006. 98, 99

Crawley, G. M., ed. (2017). *Energy Storage*. Singapore: World Scientific. DOI: 10.1142/10420.

da Rosa, A. V. (2013). *Fundamentals of Renewable Energy Processes*, 3rd ed. Oxford: Academic Press. DOI: 10.1016/B978-0-12-397219-4.00017-5.

Deffeyes, K. S. (2005). *Beyond Oil: The View from Hubbert's Peak*, New York: Hill and Wang.

Deffeyes, K. S. (2001). *Hubbert's Peak: The Impending World Oil Shortage*, Princeton, NJ: Princeton University Press.

Diamond, R. C., Ye, Q., Feng, W., Yan, T., Mao, H. W., Li, Y. T., Guo, Y. C., and Wang, J. L. (2013). "Sustainable building in China—A green leap forward?" *Buildings* 3, pp. 639–658. DOI: 10.3390/buildings3030639. 96

Dunlap, R. A. (2019). *Sustainable Energy*, 2nd ed. Boston, MA: Cengage. 87

Fay, J. A. and Golomb, D. S. (2012). *Energy and the Environment, Scientific and Technological Principles*, 2nd ed. New York: Oxford University Press. DOI: 10.1002/ep.11611.

Houghton, J. (2009). *Global Warming: The Complete Briefing*, 4th ed. Cambridge: Cambridge University Press. DOI: 10.1017/CBO9780511841590.

Huggins, R. (2016). *Energy Storage: Fundamentals, Materials and Applications*, 2nd ed. Berlin: Springer. DOI: 10.1007/978-3-319-21239-5.

Kaltschmitt, M., Streicher, W., and Wiese, A. (2007). *Renewable Energy: Technology, Economics and Environment*. Berlin: Springer. DOI: 10.1007/3-540-70949-5.

Kitchen, D. E. (2014). *Global Climate Change: Turning Knowledge into Action*. Boston, MA: Pearson.

Laherrère, J. (2001). "Estimates of oil reserves." Paper presented at the *EMF/IEA/IEW Meeting*, Laxenburg, Austria. June 19, 2001. 23

Maggio, G. and Cacciola, G. (2012). "When will oil, natural gas and coal peak?" *Fuel* 98, pp. 111–123. DOI: 10.1016/j.fuel.2012.03.021. 21, 22

Masters, G. M. (2013). *Renewable and Efficient Electric Power Systems*, 2nd ed. Hoboken, NJ: John Wiley & Sons.

Meinshausen, M., Raper, S. C. B., and Wigley, T. M. L. (2011). "Emulating coupled atmosphere-ocean and carbon cycle models with a simpler model, MAGICC6 - Part 1: Model description and calibration," *Atmospheric Chemistry and Physics* 11, pp. 1417–1456. DOI: 10.5194/acp-11-1417-2011. 37

Ngô, C. and Natowitz, J. B. (2009). *Our Energy Future*. Hoboken, NJ: Wiley. DOI: 10.1002/9780470473795.

Nugent, D. and Sovacool, B. K. (2014). "Assessing the lifecycle greenhouse gas emissions from solar PV and wind energy: A critical meta-survey," *Energy Policy* 65, pp. 229-244. DOI: 10.1016/j.enpol.2013.10.048.

Patel, M. R. (2006). *Wind and Solar Power Systems: Design, Analysis and Operation*, 2nd ed. Boca Raton, FL: CRC Press. DOI: 10.1201/9781420039924.

Peake, S. and Smith, J. (2009). *Climate Change: From Science to Sustainability*. Oxford: Oxford University Press.

Sovacool, B. K. (2009). "Contextualizing avian mortality: A preliminary appraisal of bird and bat fatalities from wind, fossil-fuel, and nuclear electricity," *Energy Policy* 37, pp. 2241–2248. DOI: 10.1016/j.enpol.2009.02.011. 56

Tester, J. W., Drake, E. M., Driscoll, M. J., Golay, M. W., and Peters, W. A. (2012). *Sustainable Energy: Choosing Among Options*, 2nd ed. Cambridge, MA: MIT Press.

Timilsina, G. R. (2013). "Biofuels in the long-run global energy supply mix for transportation," *Philosophical Transactions of the Royal Society. A* 372 20120323. DOI: 10.1098/rsta.2012.0323.

Vanek, F. M., Albright, L. D., and Angenent, L. T. (2016). *Energy Systems Engineering: Evaluation and Implementation*, 3rd ed. New York: McGraw Hill.

Wagner, T., Themeßl, M., Schüppel, A., Gobiet, A., Stigler, H., and Birk, S. (2017). "Impacts of climate change on stream flow and hydro power generation in the Alpine region," *Environmental Earth Sciences* 76. 4. DOI: 10.1007/s12665-016-6318-6. 101

Wang, M., Han, J. W., Dunn, J. B., Cai, H., and Elgowainy, A. (2012). "Well-to-wheels energy use and greenhouse gas emissions of ethanol from corn, sugarcane and cellulosic biomass or US use" *Environmental Research Letters.* 7, 045905. DOI: 10.1088/1748-9326/7/4/045905. 85

Wolfson, R. (2012). *Energy, Environment, and Climate*, 2nd ed. New York: Norton.

Wu, Z. J., Dou, X. B., Chu, J. W., and Hu, M. Q. (2013). "Operation and control of a direct-driven PMSG-based wind turbine system with an auxiliary parallel grid-side converter," *Energies* 6, pp. 3405–3421. DOI: 10.3390/en6073405.

Yasmeena, S. and Tulasiram, Das G. (2015). "A review of technical issues for grid connected renewable energy sources," *International Journal of Energy and Power Engineering. Special Issue: Energy Systems and Developments* 4, pp. 32–32. DOI: 10.11648/j.ijepe.s.2015040501.14.

Renewable Energy

Volume 2: Mechanical and Thermal Energy Storage Methods

Richard A. Dunlap
Dalhousie University

SYNTHESIS LECTURES ON RENEWABLE ENERGY TECHNOLOGIES #6

ABSTRACT

The present book reviews the possible methods of storing energy in the form of mechanical or thermal energy. The methods for mechanical energy storage that are reviewed include those that make use of gravitational potential energy, such as pumped hydroelectric energy storage and the potential energy associated with solid masses. The storage of energy by compressing air and the storage of energy as rotational energy in a flywheel are also described. The applications of these methods include grid storage and grid stabilization, as well as energy storage for transportation. There are three basic categories of thermal storage are (1) sensible heat of materials, (2) the latent heat of phase transitions and (3) the heat associated with chemical reactions. Sensible heat storage is discussed in terms of its application to residential heating, community-based heat storage, solar ponds, and thermal storage for grid-integrated energy systems. Latent heat storage may be used for both heating and cooling purposes and examples utilizing the solid-liquid transition of water and the liquid-gas transition of air are described. The basics of energy storage utilizing thermochemical reactions are presented.

KEYWORDS

renewable energy, sustainability, energy storage technology, climate change, alternative energy, hydroelectric energy storage, grid storage

Contents

Preface

At present, approximately 80% of our energy worldwide comes from the combustion of fossil fuels. This approach to energy is not sustainable because of the limited fossil fuel resources available. As well, the need to change to non-fossil fuel energy sources is accentuated by the adverse environmental effects of continued fossil fuel use. Most notable of the environmental consequences of fossil fuel use is global climate change. Although the transition to renewable carbon-free energy sources is essential, it is not easy. A significant aspect of the use of renewable energy sources is the need for energy storage. Most renewable energy sources are neither constant in time, nor are they readily portable. These two features are a requirement for much of our energy use. Specifically, a reliable supply of heat and electricity is needed for residential, as well as commercial and industrial needs, and a portable source of energy is essential for most transportation applications.

The present volume considers some of the important technologies for energy storage that utilize mechanical methods and thermal methods to store energy. Chapter 1 considers mechanical energy storage methods. These methods include pumped hydroelectric energy storage, gravitational potential of solid masses, and flywheels. Chapter 2 considers various methods of storing thermal energy. These include the use of the sensible heat of materials, the latent heat associated with phase transitions, and heat associated with chemical reactions.

CHAPTER 1

Mechanical Energy Storage

1.1 INTRODUCTION

Mechanical energy storage methods include several diverse techniques. These are used primarily for the storage of electrical energy. This chapter reviews techniques which convert electricity into gravitational potential energy, rotational kinetic energy, or gas pressure for storage. In the first case, electricity is used to move either a liquid mass (water) or a solid mass vertically in a gravitational field in order to store energy. In the second case, electricity is used to rotate a mass to store energy; and in the third case, electricity is used to pressurize a gas.

1.2 PUMPED HYDROELECTRIC

In a conventional pumped hydroelectric storage facility, water flows between an upper reservoir and a lower water supply (reservoir, river, lake, or ocean) where the upper reservoir is supplied only by water pumped from the lower reservoir. The reservoirs may be natural or artificial (anthropogenic). The most common configuration uses an artificial upper reservoir and a natural water source (e.g., river) as the lower reservoir. If the lower reservoir is connected to a natural water source, then the system is referred to as an open-loop system. If both reservoirs are artificial and have no sources of water other than that which is cycled through the pumps/turbines, then the system is referred to as a closed-loop system. These types of systems are illustrated in Figure 1.1.

Overall storage efficiency is limited by the motor/pump efficiency, turbine/generator efficiency, and water loss in upper reservoir due to evaporation. Net efficiencies are typically in the 70–80% range (compared to conventional hydroelectric generating facilities which operate in the 85–90% efficiency range). Since the storage and recovery of electrical energy requires both pumping (to store the energy) and generation (to recover the energy), the overall efficiency is the result of these combined processes.

Pumped hydroelectric storage is a commonly used method of topping up grid electricity during times of higher demand. It is a common method of load leveling—storing energy during periods of low demand and recovering this energy during periods of high demand. Pumped hydroelectric storage is used to store electricity which has been generated by any method, not just hydroelectricity or other renewable methods. It is a convenient means of grid storage as facilities have substantial generating capacity (power), as well as considerable total energy storage capacity. In addition, this power can be brought on-line quickly to satisfy demand.

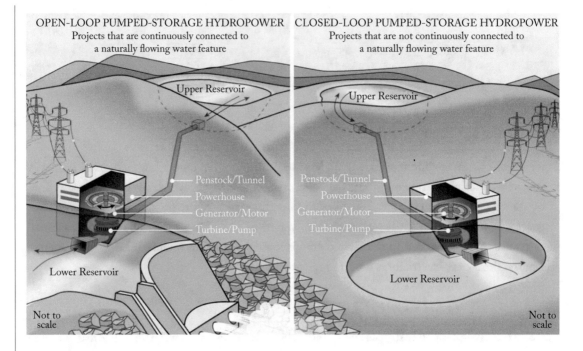

Figure 1.1: Differences between open-loop and closed-loop pumped hydroelectric storage systems. Based on https://www.energy.gov/eere/water/pumped-storage-hydropower.

The typical design of a pumped hydroelectric facility is illustrated in Figure 1.2. Water is pumped from the lower reservoir through the penstock to the upper reservoir to store energy using the motor/pump. Electrical energy is recovered when water from the upper reservoir flows through the penstock to the turbine/generator near the lower reservoir. If the average head between the upper reservoir and the generator is h, then the total energy available from the gravitational potential of the water in the upper reservoir is

$$E = mgh = \rho g h V. \tag{1.1}$$

Here, m is the total mass of water in the upper reservoir, g is the gravitational acceleration, ρ is the density of the water (kg/m^3), and V is the total volume of the upper reservoir (m^3). When h is in meters then the energy is in Joules. Including the net system efficiency, η, this may be written as

$$E = \eta \rho g h V. \tag{1.2}$$

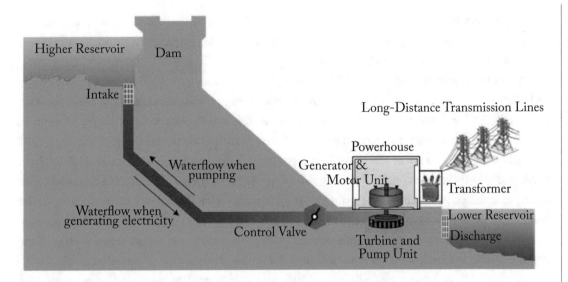

Figure 1.2: Diagram of a typical pumped hydroelectric storage facility. Based on Luo et al., 2015.

The power generated by the turbine, in Watts, is

$$P = \frac{dE}{dt} = \eta \rho g h \frac{dV}{dt} = \eta \rho g h \varphi, \tag{1.3}$$

where φ is the flow rate in m^3/s. If the total penstock cross sectional area is A, in m^2, then the flow rate is given in terms of the water velocity in the penstock, v, as

$$\varphi = vA. \tag{1.4}$$

Thus, Equation (1.3) may be written as

$$P = \eta \rho g h v A. \tag{1.5}$$

The above equations can also be used to estimate the total time, t, the facility can provide maximum power. Since $P = E/t$ then $t = E/P$ or from Equations (1.2) and (1.5),

$$t = \frac{V}{vA}, \tag{1.6}$$

where, using the SI units above, the time is given in seconds. The power is frequently expressed in MW and the energy capacity in MWh where MWh = (Joules)/(3.6×10^9).

An analysis of Equations (1.2), (1.5), and (1.6) provides the relevant characteristics of a pumped hydroelectric storage facility, that is, the total energy storage capacity, the maximum power available and the duration of maximum power. In Equation (1.2), η is a function of the motor/pump and turbine/generator design, while ρ and g are constants (ρ = 1000 kg/m^3 for fresh water and 1025 kg/m^3 for sea water, g = 9.8 m/s^2). The total energy stored is, therefore, a linear function of the average head and the upper reservoir volume. The maximum power available, as given in Equation (1.5), is a linear function of the water velocity in the penstock and the total penstock cross sectional area. The velocity of the water in the penstock is, for practical purposes, limited to about 6 m/s and penstocks are typically set at a grade of about 100% (i.e., 45° from the horizontal). This means that the penstock area is the primary design feature which determines the maximum power available. Equation (1.6) follows directly from a consideration of the volume of water and the flow rate, independent of considerations of electrical power generation, and shows that the ratio of the upper reservoir volume to the penstock area will determine the duration of power available.

Figure 1.3 shows a typical artificial reservoir that is used as the upper water source for a pumped hydroelectric storage system, i.e., Dlouhé Stráně Pumped Hydroelectric Plant in the Czech Republic. Figure 1.4 shows the same reservoir when it has been drained for maintenance, to illustrate the details of the artificial structure. An aerial view of the Ludington Pumped Hydroelectric facility in Michigan is shown in Figure 1.5. This facility was constructed during the period of 1969–1973 and, at a capacity of 1,872 MW, it was the world's largest pumped hydroelectric storage facility at the time. It is now the fifth largest pumped hydroelectric facility. Figure 1.6 shows the station under construction. The tops of the six 362 MW Francis pump/turbines are seen in the figure. Figure 1.7 shows the pump/turbine facility at the Raccoon Mountain Pumped Hydroelectric Facility in Tennessee. The facility consists of four 413 MW generators for a total generating capacity of 1,652 MW.

Figure 1.3: Upper reservoir of the Dlouhé Stráně Pumped Hydroelectric Plant in the Czech Republic; https://www.shutterstock.com/image-photo/dlouhe-strane-czech-republic-september-18-190815443?src=mokTTh2RkHqnkaFZ3cmA4A-8-67.

Figure 1.4: Upper reservoir of the Dlouhé Stráně Pumped Hydroelectric Plant in the Czech Republic. The reservoir has been drained for maintenance. Water intake for the turbines is seen in the left part of the photograph; https://www.shutterstock.com/image-photo/empty-upper-dam-pumping-hy-droelectric-power-683408029?src=mokTTh2RkHqnkaFZ3cmA4A-1-59.

Figure 1.5: Aerial view of the 1872 MW pumped hydroelectric storage facility in Ludington, MI; https://www.shutterstock.com/image-photo/pumped-storage-power-station-ludington-usa-1231207114?src=mokTTh2RkHqnkaFZ3cmA4A-1-91.

Figure 1.6: The Ludington Pumped Hydroelectric facility under construction; https://en.wikipedia. org/wiki/Ludington_Pumped_Storage_Power_Plant#/media/File:GENERATING_STATION_OF_ CONSUMER_POWER_PLANT_IN_LUDINGTON_-_NARA_-_547129_--_color_graded.tif.

Figure 1.7: Generator facility at the Raccoon Mountain Pumped Hydroelectric Storage Facility in Tennessee; https://www.tva.gov/Energy/Our-Power-System/Hydroelectric/Raccoon-Mountain.

In many cases, water for pumped hydroelectric facilities is fed through penstocks that are located underground, as in the case of the Ludington and Raccoon Mountain facilities described above. In other cases, above ground penstocks are utilized, as for the Tumut 3 generating station in Figure 1.8. This station has an average head of 151 m and six 300 MW turbines for a total capacity of 1,800 MW. Three of the six turbines can be operated as pumps for pumped hydroelectric storage. This is a pump-back facility as described below.

There are basically three different configurations that are used for pumped hydroelectric storage, depending on the number of turbines, pumps, motors, and generators that are used. These are *binary set*, *ternary set*, and *quaternary set*, as described below.

- The binary set configuration utilizes one combined pump/turbine and one combined motor/generator unit on the same shaft. This is by far the most common design, mainly because it is the most cost effective. However, it is not the most efficient, as the pump/turbine design is inevitably a compromise between the optimal design for these two purposes.

- The ternary set configuration uses a pump, motor/generator, and turbine mounted on the same shaft. The use of a separate turbine and generator allows for the performance of both the pump and turbine to be optimized.

- Quaternary set configuration uses a motor–pump assembly mounted on one shaft and a turbine-generator assembly mounted on another shaft. The two units operate independently and are, in general, located in separate powerhouses.

Figure 1.8: Tumut 3 hydroelectric generating station in New South Wales, Australia showing penstocks; https://commons.wikimedia.org/wiki/File:Tumut3GeneratingStation.jpg.

1.2.1   PUMP-BACK HYDROELECTRIC STORAGE

Pumped hydroelectric storage that is incorporated into a conventional hydroelectric facility is referred to as a pump-back facility. In this case, an upper reservoir, formed by constructing a dam across a river, can be augmented in times of low demand by pumping water into the reservoir from the river below the dam.

The Grand Coulee Dam on the Columbia River in Washington State is a notable example of a hydroelectric generating station with a pump-back facility. The Grand Coulee Dam was initially designed in 1935 as a means of irrigation control; however, in 1941 it became operational as a hydroelectric facility. Over the years the conventional hydroelectric generating capacity has been increased and is currently at 6,495 MW supplied by 27 Francis turbine-generators. Between 1973 and 1984, six motor-pump/turbine-generator units were installed for a total pump-back capacity of 314 MW. This pump-back facility is named the John W. Keys III Pump-Generating Plant and a diagram of this plant is shown in Figure 1.9. The pump-back plant cycles water between the

Franklin D. Roosevelt Lake on the downstream side of the dam and Banks Lake on the upstream side of the dam with a head of 85 m.

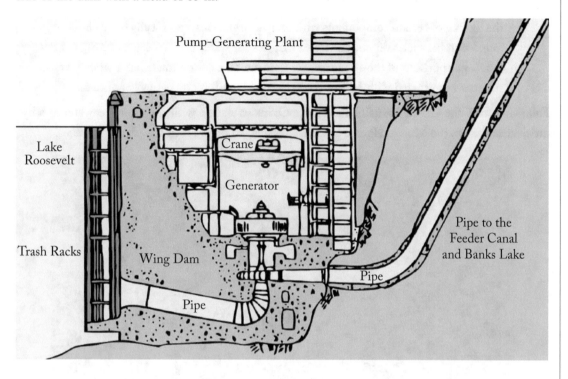

Figure 1.9: Diagram of the pump-back facility at the Grand Coulee Hydroelectric Plant in Washington. Based on https://www.usbr.gov/pn/grandcoulee/pubs/powergeneration.pdf.

1.2.2   SEAWATER-BASED PUMPED HYDROELECTRIC STORAGE

In principle, the source of water for a pumped hydroelectric storage plant can be the ocean rather than a river, lake, or reservoir. To date, one pumped hydroelectric storage facility that utilizes seawater has been constructed. The Okinawa Yanbaru Seawater Pumped Storage Power Station in Kunigami, Japan, as shown in Figure 1.10, was a 30 MW facility that was operational between 1999 and 2016. As shown in the figure, the upper reservoir was artificially constructed. The details of the intake/outlet in the Philippine Sea are shown in Figure 1.11. While the Okinawa station has demonstrated the possibility of using seawater in a pumped hydroelectric storage plant, it also emphasized some of the difficulties of using seawater in a hydroelectric facility. These difficulties include:

- the corrosive properties of seawater (compared with fresh water). This problem required certain steel components that are in contact with the seawater to be replaced with stainless steel components. The use of cathodic protection on some of the equipment was also found to be important; and

• the fouling of certain components with marine organisms, specifically barnacles. It was found that barnacles do not adhere to surfaces for water flow rates greater than about 5 m/s. For portions of the system where flow rates are substantially less than this value, surfaces were coated with hydrophobic paint, which inhibits barnacle adhesion.

Following from the experience gained from the Okinawa station with the use of seawater in a hydroelectric facility, the Muuga Harbor project, as discussed below, is in the planning stages.

Figure 1.10: Aerial photograph of the Okinawa Yanbaru Seawater Pumped Storage Power Station showing the octagonal artificial reservoir (Fujihara, Imano, and Oshima, 1998; Photograph: Roger Dargaville/Agency of Natural Resources and Energy Japan).

Figure 1.11: Seawater intake/outlet on the Philippine Sea of the Okinawa Yanbaru Seawater Pumped Storage Power Station; https://commons.wikimedia.org/wiki/File:Seawater_intake-outlet_of_Yanbaru_Power_Station.jpg.

1.2.3   SUB-SURFACE PUMPED HYDROELECTRIC STORAGE

In addition to the designs previously described the necessary head between two water sources can be obtained by using a surface-level upper reservoir and an underground lower reservoir. The general concept of the sub-surface pumped hydroelectric storage facility is illustrated in Figure 1.12. While this approach may, in general, appear to be more complex than the traditional elevated upper reservoir design, the possibility of utilizing abandoned mines as the lower reservoir may actually make this design more cost effective. In addition, the depth of many possible mine sites would provide a head that is considerably greater than that associated with traditional pumped hydroelectric stations. For example, a possible facility in the Pyhäsalmi Mine, in the town of Pyhäjärvi, Finland, which is scheduled to cease mining operations in 2019, would have a head of 1,450 m. While no facilities of this type have been constructed, a number of abandoned mine locations have been considered for possible future projects. A future renewable energy economy would make available a significant number of former coal mines that would be suitable locations for pumped hydroelectric storage.

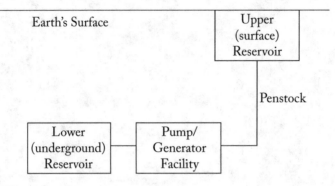

Figure 1.12: General design of an underground (sub-surface) pumped hydroelectric storage facility.

1.2.4   THE MUUGA SEAWATER PUMPED HYDROELECTRIC STORAGE PROJECT

The planned energy storage facility at Muuga Harbour, Estonia (about 13 km northeast of the capital city of Tallinn) makes use of both the concept of seawater pumped hydroelectric storage and sub-surface pumped hydroelectric storage. This project utilizes seawater, obtained from Muuga Harbour as the upper reservoir and a cavern excavated in granite bedrock as the lower reservoir. The general layout of the facility is shown in Figure 1.13. From the intake, seawater travels about 1.7 km southeast through the 7-m dimeter headrace (penstock) to the underground reservoir. A vertical section of the facility is illustrated in the lower part of the figure and shows the location of the turbine hall at a depth of 530–565 m below the surface. Energy is stored by pumping water out of the underground reservoir and into the ocean. Air enters through the ventilation shafts shown in the lower part of the figure. Energy is recovered by allowing seawater to flow from the ocean to the underground reservoir.

The planned turbine facility will consist of one 100 MW and two 175 MW reversible vertical-shaft Francis pump/turbines and one 50 MW vertical-shaft Francis turbine. The total generating capacity will be 500 MW and nominal output can be maintained for a period of about 12 hr.

Design of this project began around 2010 and the facility was intended to be operational by 2020. However, construction has been delayed because of difficulties related to obtaining the necessary municipal government approvals. When constructed, the Muuga Harbour pumped hydroelectric station will represent a novel approach to pumped hydroelectric storage that eliminates many geographical concerns for developing new energy storage facilities.

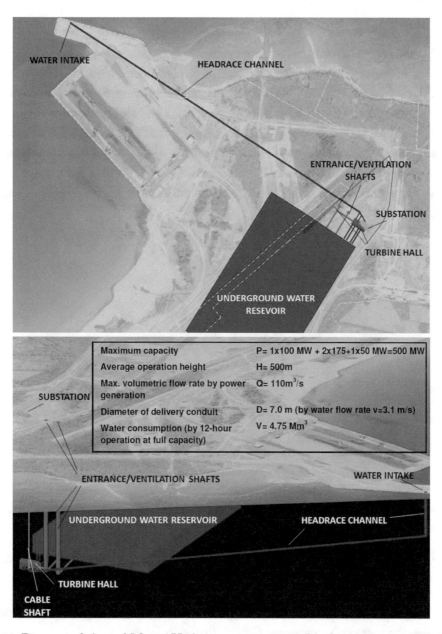

Maximum capacity	P= 1x100 MW + 2x175+1x50 MW=500 MW
Average operation height	H= 500m
Max. volumetric flow rate by power generation	Q= 110m³/s
Diameter of delivery conduit	D= 7.0 m (by water flow rate v=3.1 m/s)
Water consumption (by 12-hour operation at full capacity)	V = 4.75 Mm³

Figure 1.13: Diagram of planned Muuga Harbour seawater pumped hydroelectric storage facility. (Top) Aerial view showing the location of the seawater intake and underground reservoir. (Bottom) Vertical cross-section showing the elevation of the seawater intake, the headrace (penstock), underground reservoir, and turbine hall. (Reproduced with permission from Energiasalv Pakri OÜ, 2010.)

1.2.5   UNDERWATER RESERVOIRS

The use of an underwater chamber as a storage reservoir is an interesting variation on the conventional concept of pumped hydroelectric storage that has been considered in recent years. A basic diagram of this concept, known as ORES (Ocean Renewable Energy Storage), is shown in Figure 1.14. The chamber, in this case a spherical concrete chamber, is located on the ocean floor at a great depth. Energy is stored when the water is pumped out of the storage volume and electricity is generated by the turbine when water is allowed back into the chamber. It is important to realize that the buoyancy of the chamber, when it is evacuated, must be counteracted in order to keep the chamber on the ocean floor. Thus, sufficient ballast must be incorporated into the design of the device for this purpose.

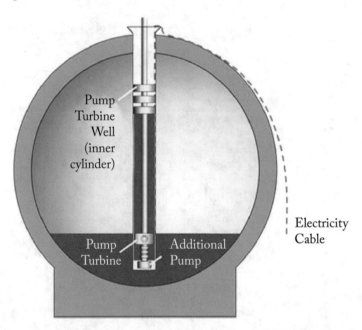

Figure 1.14: Diagram of the ORES concept. Based on Hahn et al., 2017, reproduced with permission from Elsevier.

Following from the mathematical description of conventional pumped hydroelectric energy, as given above, the energy capacity of the ORES storage chamber is

$$E = \eta \rho g d V, \tag{1.7}$$

where, as before, η is the net system efficiency, ρ is the density, in this case the density of sea water, 1,025 kg/m^3, and g is the gravitational acceleration. In Equation (1.7), d is the depth of the storage chamber in the ocean and V is the inner volume of the chamber.

From a practical standpoint, the volume of the storage chamber will be substantially less than the volume of the upper reservoir in a conventional pumped hydroelectric storage plant. On the other hand, the water depth may be greater than the typical head in the conventional system. Overall, however, one might expect the total storage capacity to be substantially less than that for existing pumped hydroelectric facilities.

Some typical design parameters of an ORES unit are given in Table 1.1. The table shows that the unit as described has a total energy storage capacity of about 15 MWh. By comparison, a typical conventional pumped hydroelectric storage facility, e.g., Raccoon Mountain, as described above, has a storage capacity of about 35 GWh. The ORES concept, however, may have some advantages over other approaches to energy storage. They may be constructed as energy storage farms consisting of multiple units, rather than a single unit, thereby increasing capacity proportionally. Since they are located underwater, they may be a convenient means of storing electricity from offshore wind farms. The proximity to the generating devices has clear advantages and the combination of offshore wind turbines with underwater storage may minimize terrestrial environmental impact. This type of storage unit is in the early experimental stages but may play a role in future renewable energy development.

Table 1.1: Typical design parameters and calculated energy storage capacity for an ORES unit

Parameter	Symbol	Value	Units
Depth	d	500	m
Sphere diameter	-	30	m
Sphere volume	V	14,137	m^3
Net efficiency	η	0.75	-
Energy storage capacity	E	14.8	MWh
Power generated	P	5	MW
Duration of power	t	2.96	h

1.2.6   GRAVITY POWER STORAGE

A novel approach to pumped hydroelectric storage is being developed by the company Gravity Power LLC. Their Gravity Power Module (GPM) consists of a vertical cylindrical shaft cut into the earth containing a piston which can move vertically in the shaft. Typically, the piston is a composite made using reclaimed rock from the excavation along with concrete. The shaft above and below the piston is filled with water. Figure 1.15 shows the principle of operation. Energy is stored by pumping water from above the piston to the space below the piston, as shown in Figure 1.15(a), thereby raising the piston. Energy is generated by allowing the piston to fall, thereby forcing water from below the piston to pass through the generator to the space above the piston. The total energy storage capacity follows from Equation (1.2) and can be written as

$$E = \eta(\rho_{piston} - \rho_{water})gV_{piston}\Delta h, \tag{1.8}$$

where η is the net efficiency, $(\rho_{piston} - \rho_{water})$ is the difference in density between the piston and the water, g is the gravitational acceleration, V_{piston} is the volume of the piston, and Δh is the change in vertical position of the center of mass of the piston. The dimensions of this device can be scaled, but a typical module might consist of a 500-m deep shaft, 30 m in diameter containing a 250-m long piston. The energy storage capacity of such a device is obtained using typical values of the parameters in Equation (1.8); $\eta = 0.8$ and $\rho_{piston} = 2,500$ kg/m³. So, for $V_{piston} = \pi(15$ m$)^2 \times 250$ m $= 1.77 \times 10^5$ m3 and $\Delta h = 250$ m, then

$$E = (0.8)\times(2,500 - 1,000)\text{kg/m}^3 \times 9.8 \text{ m/s}^2 \times 1.77 \times 10^5 \text{ m}^3 \times 250 \text{ m} = 5.21 \times 10^{11} \text{ J}, \tag{1.9}$$

or 145 MWh. This can represent a maximum power of 40 MW for a duration of about 3.6 hr.

　　While a full-scale device of this type has not been constructed, an MW-size demonstration project under development.

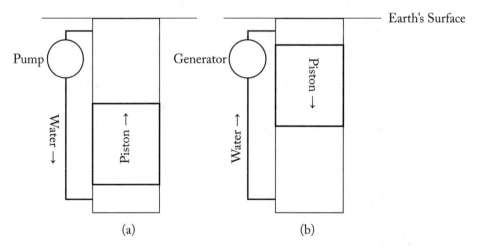

(a)   (b)

Figure 1.15: Concept of the Gravity Power Module: (a) energy storage and (b) energy generation.

1.2.7   WORLD USE OF PUMPED HYDROELECTRIC STORAGE

While several of the possibilities previously mentioned are still in the early development stages, traditional pumped hydroelectric storage has made substantial contributions to grid storage and, at present, represents the most highly developed energy storage method for this purpose. While infrastructure costs of pumped hydroelectric storage can be quite high, maintenance is generally minimal and the lifetime of facilities is quite long. The world's largest pumped hydroelectric storage facilities, based on maximum generating capacity, are summarized in Table 1.2. As the table shows the largest facilities have generating capacities of 2–3 GW, which is large compared to coal-fired and nuclear generating stations, which are typically around 1 GW.

Table 1.2: The largest, by generating capacity, pumped hydroelectric storage stations in the world

Station	Country	Capacity (MW)
Bath County Pumped Storage Station	Virginia, U.S.	3,003
Guangdong Pumped Storage Power Station	Guangdong, China	2,400
Huizhou Pumped Storage Power Station	Guangdong, China	2,400
Okutataragi Pumped Storage Power Station	Hyōgo, Japan	1,932
Ludington Pumped Storage Power Plant	Michigan, U.S.	1,872

The distribution of pumped hydroelectric storage plants world wide is summarized in Table 1.3. It is clear from the table that East Asia (which includes China) and Europe have the greatest storage capacity. In fact, in Europe, pumped hydroelectric storage represents about 5.5% of base load capacity. In North America it represents about 2.5% of base load capacity.

Table 1.3: Breakdown of geographical distribution of pumped hydroelectric storage facilities, as of 2015. Data adapted from U.S. Department of Energy, Global Energy Database

Region	Total Capacity (MW)
East Asia	57,999
Europe	50,949
North America	22,618
South and Central Asia	6146
Southeast Asia and Pacific	2425
Middle East and North Africa	1744
Africa	1580
Latin America and Caribbean	1004
World Total	144,465

The distribution of pumped hydroelectric storage facilities in the U.S. is illustrated on the map in Figure 1.16. It is seen that the majority of facilities are located along the east coast (near the Appalachian Mountains) and in California. It is also seen that nearly all the facilities are open-loop systems; only two closed-loop systems exist in California.

Licensed Pumped Storage Projects

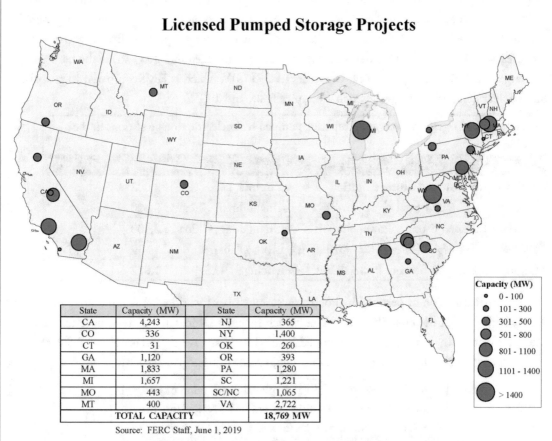

State	Capacity (MW)	State	Capacity (MW)
CA	4,243	NJ	365
CO	336	NY	1,400
CT	31	OK	260
GA	1,120	OR	393
MA	1,833	PA	1,280
MI	1,657	SC	1,221
MO	443	SC/NC	1,065
MT	400	VA	2,722
TOTAL CAPACITY			**18,769 MW**

Source: FERC Staff, June 1, 2019

Capacity (MW)
- 0 - 100
- 101 - 300
- 301 - 500
- 501 - 800
- 801 - 1100
- 1101 - 1400
- > 1400

Figure 1.16: Map of pumped hydroelectric storage facilities in the U.S. showing the generating capacity of each plant; https://www.ferc.gov/industries/hydropower/gen-info/licensing/pump-storage/licensed-projects.pdf?csrt=16406207306755896610.

1.3   GRAVITATIONAL POTENTIAL OF SOLID MASSES

Following along the lines of pumped hydroelectric storage, it has been suggested that solid masses may be raised and lowered to store energy by making use of the differences in gravitational potential. The energy storage capacity is described by

$$E = \eta mg \Delta h, \tag{1.10}$$

where η is the net system efficiency, m is the mass, g is the gravitational acceleration, and Δh is the difference in height. It is clear that the energy storage capacity is maximized by maximizing the mass and the change in height, although, from a practical standpoint, there may be limits to either or both of these quantities. One approach is to maximize the height difference but to utilize moderate masses. One suggestion along these lines is the use of masses raised and lowered from an ocean barge in deep water, where the ocean depth, e.g., 4,000 m, provides a large value of Δh. It should be noted in this case, however, that the mass is not the total mass, but the difference between the mass and the water displaced, as in Equation (1.8). Another approach along similar lines is to raise and lower masses from a buoyant platform floating in the atmosphere. A height of 20,000 m would provide a large value of Δh and solar photovoltaic panels on the upper surface of the platform could provide the electricity to be stored.

Perhaps the most practical approach to raising and lowering solid masses to store energy is a rail-based system for moving masses up and down a long incline. Advance Rail Energy Storage (ARES) LLC has tested a prototype system of this type in Tehachapi, California (near a major wind farm) as shown in Figure 1.17. The rail vehicle containing an electric motor/generator is loaded with large masses and travels along a track. Additional non-motorized cars may be attached to the motorized vehicle to increase mass. Typical masses in the order of 200 t would be used. Electrical connection for the motor/generator is provided by a third rail, as is utilized for electric trains.

Figure 1.17: Prototype rail energy storage vehicle in Tehachapi, California. Note the wind farm along the horizon (Cava et al., 2016).

As with pumped hydroelectric storage facilities, appropriate geography is necessary for a rail energy storage system. A grade of less than about 10% is needed for the rail drive system to function, so a long gradual incline is necessary. An incline of 8% over a distance of 15 km would give a total difference in height of 1200 m, corresponding to a stored energy of about 0.5 MWh per (200 t) vehicle. Note that the necessary geography is quite different than that required for pumped hydroelectric storage where typically a 100% grade is used for penstocks.

Such a concept is scalable, within a fairly wide range of sizes. Capacity might range from about 25 MW with 6 MWh storage (for an output duration of about 15 min) to 2,000 MW with 240,000 MWh storage (for an output duration of 120 hr).

Although such an approach to energy storage is still at its early stages of technical development, there are a number of possible advantages over the commonly used pumped hydroelectric storage system. These include:

- about half the initial infrastructure cost,

- scalable over a larger range of sizes,

- potentially greater availability of locations with appropriate geography,

- minimal post-lifetime environmental impact, and

- excellent net efficiency of around 90%.

1.4 COMPRESSED AIR ENERGY STORAGE

The basic concept of compressed air energy storage is quite simple. Electricity is used to operate a motor-pump to compress air in a confined volume. The air is then expended through a turbine, which turns a generator to recover the stored electricity. However, in practice the process is not so simple, and not so easy. In order to understand how compressed air energy storage actually works, we need to look in some detail at the thermodynamics of compressing a gas.

1.4.1 PHYSICS OF COMPRESSED AIR ENERGY STORAGE

We consider the adiabatic compression of a gas. In an adiabatic process the system is thermally isolated from its surroundings and no heat is transferred. The first law of thermodynamics gives the change in total internal energy of a gas, ΔU, in terms of the heat supplied to the system, ΔQ, and the work done by the system, W, as

$$\Delta U = \Delta Q - W, \tag{1.11}$$

where the work done by the system may be written as

$$W = P\Delta V. \tag{1.12}$$

For an adiabatic process, $\Delta Q = 0$, which gives

$$\Delta U = - P\Delta V. \tag{1.13}$$

For an ideal gas the internal energy is given by

$$U = c_v nT, \tag{1.14}$$

where c_v is the molar heat capacity, n is the number of moles and T is the temperature. Combining Equations (1.13) and (1.14) gives

$$c_v n\Delta T = - P\Delta V. \tag{1.15}$$

For compression, $\Delta V < 0$ and the change in temperature will be $\Delta T > 0$, i.e., heating during compression. For expansion $\Delta V > 0$, $\Delta T < 0$ and the gas will cool. In terms of compressed air energy storage, the work performed by the compressor both compresses the gas and heats it. When energy is recovered by expanding the gas (to turn a turbine), the gas cools and all of the energy input into the system is returned (except for losses due to friction etc. in the machinery). The temperature of the gas after compression, T_a, relative to the temperature of the gas before compression, T_b, may be related to the pressures before, P_b, and after, P_a, compression as

$$T_a = T_b \left(\frac{P_a}{P_b}\right)^\chi, \tag{1.16}$$

where the exponent χ is related to the ratio of specific heats at constant pressure, c_P, and at constant volume, cV_P, as

$$\chi = \left(\frac{c_P}{c_V} - \frac{c_V}{c_P}\right). \tag{1.17}$$

For dry air, $\chi \approx 0.29$. In practice, this means that the temperature of the compressed gas can be quite high, and this may cause practical problems.

   In practice, compression of a gas can be adiabatic, diabatic, or isothermal. As noted above, adiabatic compression refers to the case where there is no heat transfer between the system and the environment, so that the system temperature is defined by Equation (1.15). In principle, a system could be designed so that the heat produced during compression was transferred to a thermal storage medium (e.g., stone, oil, salt, etc.; see Chapter 2). In this case the stored heat can be returned to the gas during the expansion phase to increase the efficiency of energy recovery.

   Diabatic compression refers to the situation where the heat associated with compression is partially removed from the gas by means of heat exchangers (referred to as intercoolers). The compressed gas remains at a temperature that is less than for the adiabatic case. In practice, this means that the gas must be heated prior to expansion. This is done using a natural gas-fired heater.

Finally, isothermal compression refers to the situation where all heat generated by the compression is transferred to the environment, thereby keeping the temperature of the compressed gas the same as before the compression.

A simple way of estimating the maximum energy storage capacity in a compressed gas is to consider the isothermal case. The gas is described by the ideal gas law

$$PV = nRT, \tag{1.18}$$

where n is the number of moles of gas and R is the ideal gas constant. The work done on the system to compress the gas from an initial pressure and volume, P_i and V_i, to a final pressure and volume, P_f and V_f, is given by

$$E = \int_{V_f}^{V_i} P dV = \int_{V_f}^{V_i} \frac{nRT}{V} dV = P_f V_f \cdot \ln \frac{P}{P_i}, \tag{1.19}$$

and is equal to the theoretical energy stored. A simple calculation shows that the energy stored in a 1 m³ volume for an initial pressure of 100 kPa (1 atmosphere) and a final pressure of (say) 10 MPa (100 atmospheres) is

$$E = (10 \text{ MPa}) \cdot \ln(10^7/10^5) = 46 \text{ MJ}, \tag{1.20}$$

or about 13 kWh. This provides a rough estimate of the maximum energy capacity in a compressed air energy storage facility.

1.4.2   LOCATIONS FOR COMPRESSED AIR ENERGY STORAGE

It is clear from the above discussion that grid-scale compressed air energy storage requires the use of considerable pressures. More importantly, however, is the need for an appropriately large volume that will withstand the required pressures. There are several possibilities for suitable locations. These fall into two basic categories: underground locations and underwater locations. The most convenient underground locations for compressed air energy storage are geological formations associated with salt domes. A salt dome is a natural formation where a column of salt intrudes upward into an overlaying layer of sedimentary rock. Figure 1.18 shows a diagram of a salt dome. Formations often have associated oil deposits, as shown in the figure. A cavern appropriate for compressed air energy storage can be created by solution-mining the salt dome. This process involves dissolving the salt by flushing the deposit with water. Caverns prepared in this manner have long been used for the storage of gases and liquids such as natural gas, hydrogen, and oil. Other possible sites include abandoned limestone mines.

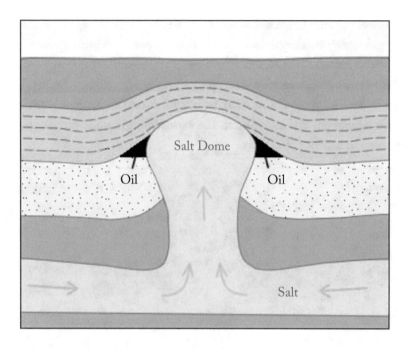

Figure 1.18: **Diagram of a salt dome. Based on** https://geology.com/stories/13/salt-domes/.

Underwater compressed air energy storage requires the use of an appropriate containment vessel. Compressed air may be stored at a pressure comparable to the hydrostatic pressure. At a depth, d, in water the hydrostatic pressure is given by

$$P = \rho g d, \tag{1.21}$$

where ρ is the water density and g is the gravitational acceleration. It is clear that the storage capacity will increase with water depth. Oceans provide the most obvious suitable locations, but deep-water lakes are also possibilities.

Vessels for underwater compressed air energy storage facilities fall into two general categories: flexible structures and rigid structures. Flexible vessels can be manufactured from air and water tight fabric. Energy is stored in the vessel by filling it with air at a pressure that is counteracted by the hydrostatic pressure of the surrounding water. Energy is recovered from the vessel by displacing the air with water. In the charged state the vessel will be subject to a buoyancy force which depends on the difference in density between the water and the compressed air ($\rho_{water} - \rho_{air}$):

$$F = (\rho_{water} - \rho_{air})gV, \tag{1.22}$$

where V is the vessel volume. The vessel must, therefore, be anchored appropriately to the sea (or lake) bottom. Rigid vessels constructed of (for example) concrete can incorporate sufficient ballast to counteract buoyant forces.

Figure 1.19 shows a schematic of a compressed air energy storage facility. The gas is compressed diabatically and, as noted above, must be heated during expansion. During expansion the gas is mixed with natural gas and combusted. The expanding gas turns the natural gas turbine to drive the generator and produce electricity.

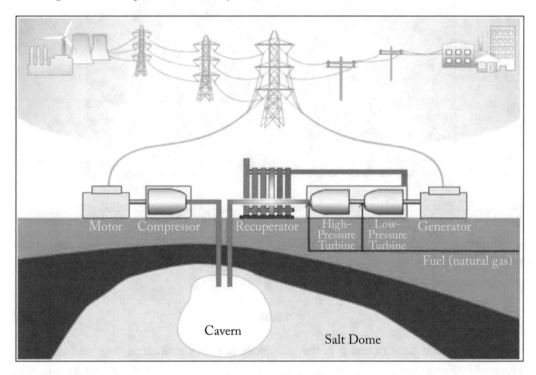

Figure 1.19: Schematic of a compressed air energy storage facility. Based on Akhil et al., 2013; https://www.sandia.gov/ess-ssl/publications/SAND2013-5131.pdf.

1.4.3 WORLD USE OF COMPRESSED AIR ENERGY STORAGE

At present, there are two grid-scale compressed air energy storage facilities worldwide. The first to become operational, in 1978, was the Huntorf Compressed Air Energy Storage facility in Germany. This facility utilizes two solution-mined salt domes with a total volume of 3.1×10^5 m³. The top of the caverns is located 650 m below the earth's surface and the bottom of the caverns is located at a depth of 800 m. The maximum cavern diameter is 60 m. The facility produces a peak output of 321 MW for a maximum of 2 hr at a net efficiency of about 42%. A photograph of the interior of the facility is shown in Figure 1.20. Details of the compressor/generator assembly are shown in the model depicted in Figure 1.21.

Figure 1.20: Photograph of the interior of the Huntorf compressed air energy storage facility. From the left side of the image, the units are high-pressure compressor, transmission, low-pressure compressor, motor/generator (see Figure 1.21); https://commons.wikimedia.org/wiki/File:Kraftwerk_Huntorf_innen.jpg.

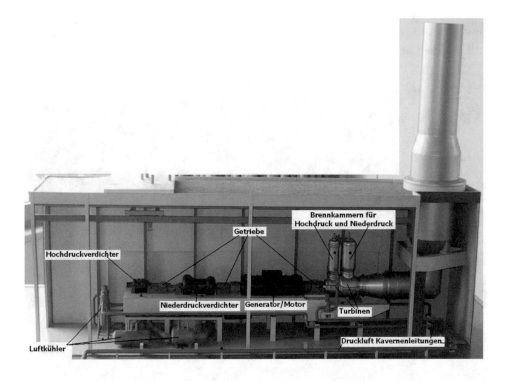

Figure 1.21: Model of the interior of the powerhouse at the Huntorf compressed air energy storage facility in Huntorf near Elsfleth in Lower Saxony, Germany (niederdruckverdichter = low-pressure compressor; hochdruckverdichter = high-pressure compressor; getriebe = transmission); https://commons.wikimedia.org/wiki/File:Kraftwerk_Huntorf_Modell.jpg.

In 1991, the world's second compressed air energy storage facility became operational in McIntosh, Alabama. This plant has a single cavern that was solution-mined from a salt dome. The top of the cavern is at a depth of 459 m and the bottom of the cavern is at a depth of 689 m. The maximum cavern diameter is 73 m and its total volume is 5.6×10^5 m^3. The facility has a peak output of 110 MW for a maximum of 26 hr. Efficiency is slightly less than for the Huntorf plant.

The design of these currently operational compressed air energy storage plants uses diabatic compression and, as a result, requires heating of the air by natural gas combustion during expansion. This approach uses fossil fuels and is, therefore, not a totally carbon-free energy storage approach. However, a detailed analysis suggests that these facilities use only 30–40% as much natural gas compared with a conventional natural gas turbine of the same energy output.

1.5   COMBINED PUMPED HYDROELECTRIC-COMPRESSED AIR ENERGY STORAGE

Another possible energy storage approach that deals with geographical constraints on traditional pumped hydroelectric storage and the need for natural gas combustion for traditional compressed air energy storage combines some aspects of both of these technologies. An illustration of a simple combined pumped hydroelectric-compressed air energy storage system is shown in Figure 1.22. To store energy, water is pumped from the open tank on the left in the figure into the sealed air-pressurized chamber on the right. Energy is recovered by allowing water to run back through the pump/turbine into the open water tank. As the water is driven by air pressure not by differences in gravitational potential, there are no constraints on geography for the construction of such a system.

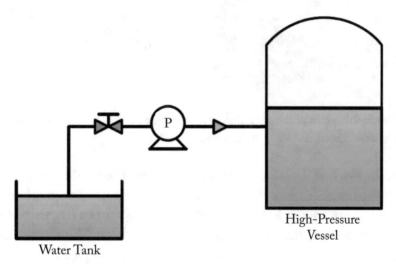

Figure 1.22: Schematic of a simple combined pumped hydroelectric-compressed air energy storage system, P = pump/turbine. Based on Yao et al., 2015.

The detailed operation of a combined pumped hydroelectric-compressed air energy storage system is described in Figure 1.23. The high-pressure vessel (5) is pre-pressurized with air by compressor (4) through valve (8). The compressor is not used subsequently in the storage or recovery of energy. To store energy, water from the tank (3) is pumped into the high-pressure vessel through the valve (7). To recover energy, water from the high-pressure vessel is forced by air pressure through valve (6) and the hydroelectric turbine (1) and back into the water tank. During energy storage and recovery, valves (9) and (10), respectively, are opened to spray water into the high-pressure vessel. This water exchanges heat with the air in the high-pressure vessel in order to reduce the effects of heating and cooling, during compression and expansion, respectively. This eliminates the need for natural gas combustion, as is used in commercial compressed air energy storage systems.

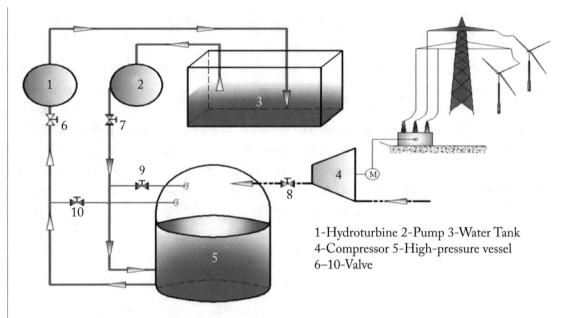

Figure 1.23: Detailed diagram of the operation of a combined pumped hydroelectric-compressed air energy storage system. Based on Wang et al., 2013.

A major drawback of the energy storage system as described is the fact that during energy storage and during energy recovery, the gas pressure in the high-pressure vessel does not remain constant. The consequence of this feature is that the energy storage rate (i.e., the power consumed) and energy recovery rate (i.e., the power generated) are not constant. This adverse characteristic of the system described can be overcome by the design of the constant pressure system as, shown in Figure 1.24. In this system both the storage vessel and the high-pressure vessel are pre-pressurized by the compressor (C1) through valves (1) and (2), respectively. Compressor C1 is not subsequently used during energy storage or recovery. Energy is stored by pumping water from the water tank to the storage vessel through valves (3) and (4). As water is pumped into this vessel, compressor (C2) transfers air from the storage vessel to the high-pressure vessel through valve (5) so as to maintain the air pressure above the water in the storage vessel at a pre-set constant value. Energy is recovered by pressurized water from the storage vessel flowing through the turbine (through valves (6) and (7)). As the water level in the storage vessel drops, compressed air is supplied from the high-pressure vessel through valve (8) in order to maintain the pressure in the storage vessel at a constant value.

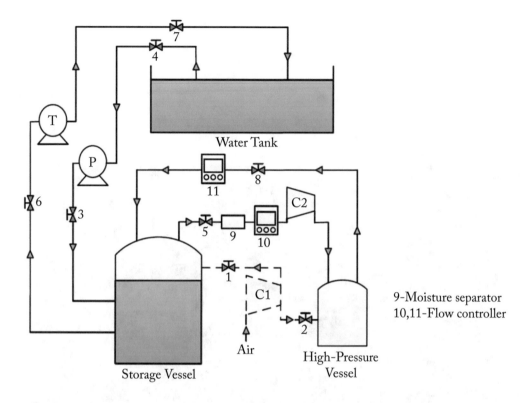

Figure 1.24: Diagram of a constant pressure combined pumped hydroelectric-compressed air energy storage system; P = pump, T = turbine. Based on Yao et al., 2015.

A test system similar to the constant pressure design described above has been constructed by Camargos et al., (2018) and is shown in the photograph in Figure 1.25. Figure 1.26 shows a diagram of the system and identifies the components. Pump (P-01) transfers water from the open water tank (TK-03) to the air/water tank (TK-02) for energy storage. A high-pressure Pelton turbine (T-01 in the figures) is used for recovery of energy from water flowing from the air/water tank (TK-02) to the open water tank (TK-03). The Pelton turbine is connected to a DC generator (GEN-01) to produce electricity. Efficiencies comparable to those of commercial compressed air energy storage facilities have been obtained. The general concept utilized in pumped hydroelectric-compressed air energy storage systems should be readily scalable for systems with different capacities and applications.

Figure 1.25: Photograph of the prototype combined pumped-hydroelectric-compressed air energy storage system. Components are described by number in Figure 1.26. Photo from Camargos et al., 2018, with permission from Elsevier.

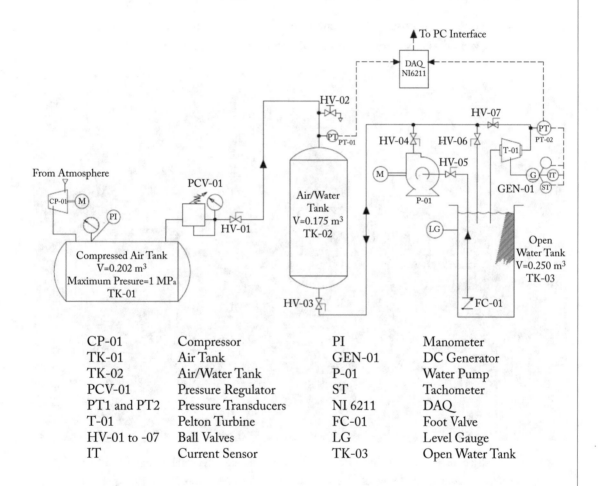

Figure 1.26: Diagram of experimental combined pumped hydroelectric-compressed air energy storage system. Based on Camargos et al., 2018, with permission from Elsevier.

1.6   FLYWHEELS

Flywheels are axially symmetric devices designed to store rotational energy and have been in use since antiquity (e.g., for potter's wheels). Applications of flywheels include the following.

- Smoothing the rotational output of devices that convert reciprocating motion to rotary motion. This was probably the first industrial use of the flywheel for steam engines developed in the late 18th century, see Figure 1.27. Flywheels on large 19th century stationary steam engines could be up to 9 m in diameter.

- Controlling or measuring the orientation of a device by use of a gyroscope. These devices were developed around the beginning of the 19th century.

Figure 1.27: 19th century steam engine with a cast iron rim-loaded flywheel. Image from Lestertair, Shutterstock.com; https://www.shutterstock.com/image-photo/munich-germany-november-26-2018-german-1250746300?src=crr_hZknlDSD0IXHnHSE4A-1-34.

More recently, flywheels have been considered as a means of storing energy. Specifically, electrical energy can be used to rotate the flywheel by means of an electric motor and the energy can be recovered by slowing the flywheel as it turns a generator. A simple system for using a flywheel to store electrical energy is shown in Figure 1.28.

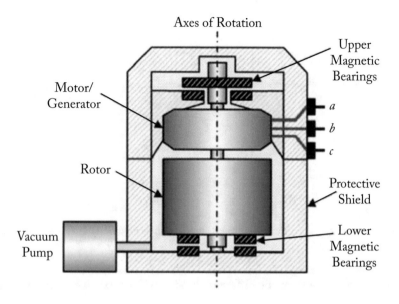

Figure 1.28: Typical design of a flywheel energy storage system. Based on Guney and Tepe, 2017, with permission from Elsevier.

1.6.1   THE PHYSICS OF FLYWHEELS

The rotational kinetic energy of an object can be calculated by beginning with the translational kinetic energy of a mass, *m*, moving with a velocity, *v*, as

$$E = \frac{1}{2}\, mv^2. \tag{1.23}$$

If we consider a differential mass, *dm*, as shown in Figure 1.29, moving with a velocity, *v*, the kinetic energy will be

$$dE = \frac{1}{2}\, v^2 dm. \tag{1.24}$$

Figure 1.29: Kinetic energy of a differential mass moving with a velocity v at a perpendicular distance r from the z-axis.

If the perpendicular distance from the differential mass to the z-axis is r, then the angular velocity about the z-axis will be

$$\omega = \frac{v}{r},\tag{1.25}$$

and Equation (1.12) may be written as

$$dE = \frac{1}{2}\,\omega^2 r^2 dm.\tag{1.26}$$

If we now consider the rotational kinetic energy of a distributed object about the z-axis, we can integrate its differential mass elements as

$$dE = \frac{1}{2}\,\omega^2 \iiint r^2 dm,\tag{1.27}$$

where the integral is over all space and is a function of the geometry and mass distribution of the object. This integral is defined as the moment of inertia of the object, I, and Equation (1.27) can be written in its usual form as

$$E = \frac{1}{2}\,I\omega^2.\tag{1.28}$$

The moment of inertia from Equation (1.27) may be written as an integral of the density distribution of the object over its volume:

$$I = \iiint \rho(r) r^2 dV,\tag{1.29}$$

where r is the vector defining the perpendicular distance between the volume element and the z-axis.

The integral in Equation (1.29) can be evaluated for masses of different geometries. In the simplest case we can consider different geometries for objects of total mass, m, that have constant density and are axially symmetric about the z-axis (axis of rotation) with a total radius, R. In this case the moment of inertia from Equation (1.29) can be expressed as

$$I = kmR^2 , \qquad (1.30)$$

where the constant k is determined by the object's geometry. Equation (1.30) can be combined with Equation (1.28) to give

$$E = \frac{1}{2} kmR^2\omega^2. \qquad (1.31)$$

Values of k for some common geometries are given in Table 1.4. For a simple flywheel, it can be seen from Equation (1.31) and the values of k in Table 1.4 that the maximum energy can be stored in a ring-shaped flywheel, as this geometry places as much of the mass as far from the axis of rotation as possible. Such a geometry is sometimes referred to as a rim-loaded flywheel, as shown in Figure 1.27. There are, however, limits to this approach for energy storage because of the internal stresses in flywheels of different geometries. From a practical standpoint, as discussed below, a thick-walled cylinder (or even a disk) is generally more appropriate than a thin rimmed cylinder. A thick-walled cylinder is defined by an inner radius of R_i and an outer radius of R_o. The moment of inertia of this geometry is given by

$$E = \frac{1}{2} m(R_i^2 + R_o^2). \qquad (1.32)$$

Table 1.4: The constant k for the moment of inertia in Equation (1.30) for some axially symmetric geometries	
Geometry	k
Solid sphere	2/5
Disk	1/2
Cylinder	1/2
Spherical shell	2/3
Ring	1

One might expect that materials such as iron, as has traditionally been used for flywheels (e.g., Figure 1.27), would be the most suitable for storing large amounts of energy. However, a careful analysis of the properties of different materials shows that this is not necessarily the case. The rim of the flywheel has the highest velocity and is, therefore, subject to the greatest stress. It can be shown that the rim stress is given by

$$\tau = \rho R^2 \omega^2, \tag{1.33}$$

where ρ is the bulk density of the material. The maximum allowable stress is the breaking stress of the flywheel material (generally taken to be the tensile strength, $\tau = \sigma$). Combining Equation (1.33) with Equation (1.31) for the energy gives the maximum energy that can be stored as

$$E = \frac{1}{2} Ik \left(\frac{\sigma}{\rho} \right), \tag{1.34}$$

where E is in J when mass is in kg, σ is in N/m^2, and ρ is in kg/m^3. The maximum energy storage capacity of a flywheel is, therefore, related to its geometry (through k), its mass, m, and the intrinsic material property (σ/ρ). Since the maximum energy is linearly related to m and (σ/ρ), choosing the material with the highest possible value of (σ/ρ) should yield the flywheel with the highest possible energy storage capacity per unit mass. Note that (σ/ρ) is in units of N $\cdot$ m/kg = J/kg. It is interesting to note that all other factors being equal, low-density materials are preferable to high-density materials.

Another way of looking at Equation (1.34) is to use the relation mass/density = volume to obtain

$$E = \frac{1}{2} k\sigma V, \tag{1.35}$$

meaning that the material with the highest tensile strength will yield the flywheel with the highest energy capacity per unit volume. Note that the units of tensile strength are N/m^2 = J/m^3.

1.6.2   FLYWHEEL DESIGN CRITERIA

While the discussion above provides some general guidance to the design of a flywheel, there are a number of specific factors that need to be considered in some detail in order to optimize its functionality and efficiency. These factors include the following:

- the flywheel material,

- the flywheel geometry,

- the flywheel enclosure,

- the motor/generator, and

- the support of the flywheel (i.e., the bearings).

Flywheel Materials

Table 1.5 gives the properties of some materials that may be considered for flywheel construction. It is clear that the use of cast iron as a flywheel material, as was traditional for steam engines in the 19th century, is not suitable for energy storage purposes. The table also shows that materials that we typically think of as "strong" are not necessarily the best choice because of their high density. Carbon fiber composites are seen to offer the best option for energy storage, both in terms of energy storage capacity per unit mass and per unit volume.

Table 1.5: Properties of some potential flywheel materials. The energy storage capacity per unit mass is related to σ/ρ and the energy storage capacity per unit volume is related to σ. Note: values are typical for each class of material, as tensile strength can vary considerable from one alloy to another

Material	Density (ρ) [kg/m^3]	Tensile Strength (σ) [N/m^2]	σ/ρ [N·m/kg]
Cast iron	7870	0.2×10^9	0.025×10^6
Magnesium alloy	1740	0.24×10^9	0.14×10^6
Steel	7870	1.72×10^9	0.22×10^6
Aluminum alloy	2700	0.59×10^9	0.22×10^6
Beryllium	1850	0.48×10^9	0.26×10^6
Titanium alloy	4500	1.22×10^9	0.27×10^6
Fiberglass	2000	1.60×10^9	0.80×10^6
Carbon fiber composite	1500	2.40×10^9	1.60×10^6

Flywheel Rotor Geometries

It is seen from the above discussion that a rim-loaded flywheel design provides the highest moment of inertia because it maximizes the distance of the mass from the axis of rotation. However, the rim of the flywheel is traveling at the greatest velocity and, therefore, has the greatest stress. Distributing some of the mass of a flywheel rotor closer to the axis of rotation (compared to a rim-loaded design), will decrease the overall moment of inertia, but it may increase the total energy storage capacity by reducing the stress at the rim and allowing greater rotational speeds. A commonly used approach to analyzing different flywheel rotor shapes is to rewrite Equations (1.34) and (1.35) in terms of the maximum energy per unit mass and per unit volume as

$$\frac{E}{m} = K\left(\frac{\sigma}{\rho}\right), \tag{1.36}$$

and

$$\frac{E}{V} = K\sigma, \tag{1.37}$$

where K is a shape factor that is determined by the rotor geometry. Figure 1.30 illustrates the values of K for some different geometries. The so-called *Disk of Laval* is the shape that has constant stress across the rotor and is also the shape that has the highest value of K. Different flywheels may utilize differently shaped rotors in order to optimize the design for a particular application. A common approach to rotor design is the use of a composite flywheel where the inner portion of the flywheel is made from a material such as steel or titanium and the rim of the flywheel, which experiences the greatest stress, is made of carbon fiber.

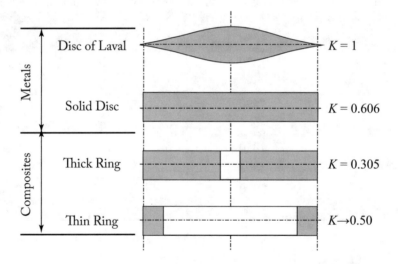

Figure 1.30: Shape factors for some different flywheel rotor geometries. Based on Sebastián and Peña-Alzola, 2012, with permission from Elsevier.

Flywheel Enclosures

In order to reduce energy loss due to friction between the flywheel rotor and air, it is customary to enclose the flywheel in a vacuum chamber, as shown in Figure 1.28. This enclosure is typically made of a structural material such as steel. In some applications where energy losses are not a major concern, flywheel enclosures may be filled with low-pressure helium.

Motor/generator

In most cases the motor/generator is mounted inside the vacuum enclosure and is coupled directly to the flywheel rotor, as shown in Figure 1.28. In some designs the motor/generator may be outside of the vacuum enclosure and may be magnetically coupled to the flywheel shaft. There are a number of different types of motor/generators that can be used for flywheel energy storage, depending on the specific application and power levels involved. One major concern that needs to be addressed with regard to the choice of the motor/generator is that of heat. Motor/generators produce heat during the charging and discharging of the flywheel. In systems which are used (for example) for backup power, the flywheel is charged or discharged only rarely, and heating is generally not an issue. In systems which are cycled often, heating can be an important factor. Heat is conducted through the shaft to the flywheel but, because the system is enclosed in a vacuum, heat is dissipated very slowly. For metallic flywheels (e.g., steel or titanium) heating is generally not a problem, however, for composite flywheels (as discussed in the applications section) excess temperatures can be detrimental to the rotor. In such cases, motor/generators which generate as little heat as possible are desirable. Permanent magnet motors are generally the best choice in such cases.

Bearings

Bearings support the flywheel rotor and allow it to rotate freely, with as little friction as possible. Generally, stand alone flywheel systems are configured with a vertical axis so that bearing loads are distributed symmetrically. Traditionally, mechanical bearings (i.e., balls in races) have been used for rotating machinery. Since flywheels that are designed to store significant amounts of energy are heavy and rotate at high speeds, bearings must be appropriately designed. In most cases, bearing life is the limiting factor for continuous flywheel operation. It should also be noted that the usual vacuum requirements for flywheel operation place additional demands on heat dissipation and lubrication.

Most advanced flywheel systems utilize magnetic bearings, sometimes referred to as active magnetic bearings (AMBs), where the flywheel rotor is levitated in a magnetic field. A common configuration uses permanent magnets to levitate the rotor assembly and small electromagnets to provide fine positioning and stability. Magnetic bearings typically introduce less energy loss than mechanical bearings. Other advantages of magnetic bearings result from the fact that the rotating flywheel assembly is not in physical contact with the stationary portions of the flywheel enclosure. This means that heat generation in the bearings is not an issue and no lubricant is needed. The lack of lubricant benefits the production of the vacuum environment. One disadvantage of magnetic bearings, however, is the need for so-called touch-down bearings, which support the flywheel in the event of a magnetic bearing failure. Magnetic bearings may also be used in conjunction with mechanical bearings in order to minimize bearing load and wear.

Superconducting levitation systems based on the Meissner effect may also be used to support a flywheel rotor. Such systems are in the early stages of commercial development.

1.6.3   APPLICATIONS OF FLYWHEELS FOR ENERGY STORAGE

Grid Storage and Stabilization

A number of commercial flywheel energy storage units are available for grid use. A typical example is shown in Figure 1.31. This unit has a storage capacity of 32 kWh and a rated power output of 8 kW, for a total duration of 4 hr. While the rated capacity is much less than pumped hydroelectric storage or proposed systems involving the gravitational potential of solid masses, flywheel energy storage modules have found applications for grid stabilization and for covering short-term power interruptions. A large number of flywheel modules can be connected together to increase energy storage and power capacity. For example, a grid storage system was installed in Stephentown, New York by Beacon Power, consisting of 200 flywheel modules for a total energy storage capacity of 5 MWh and a maximum power output of 20 MW. Typical units have a self-discharge rate of about 5% per day but can be kept fully charged from the grid until needed.

Figure 1.31: **Commercial flywheel energy storage module;** https://www.energystoragenetworks.com/hawaiian-electric-amber-kinetics-demo-kinetic-energy-storage/. **Photo courtesy of Amber Kinetics** (www.amberkinetics.com).

Vehicle Propulsion

Flywheels may provide an alternative means of vehicle propulsion. Here we consider the suitability of flywheel energy storage capacity for such an application. We consider the maximum specific energy storage for a carbon fiber composite flywheel as given in Table 1.5 of 1.6 MJ/kg (although one may choose a smaller value for safety). If we assume a total flywheel mass of 100 kg, then the total energy storage will be 160 MJ. A mid-size passenger vehicle typically requires an average of about 0.55 MJ of energy at the wheels per km of distance traveled leading to a range for the flywheel vehicle of (160 MJ/0.55 MJ/km) = 290 km, similar to that of many battery electric vehicles. In the transportation sector, flywheels have most commonly been used as a means of propulsion for city buses. In this case, weight, size, and/or cost considerations may not be as important as for personal passenger vehicles.

It may also be possible to construct hybrid vehicles which use a flywheel in conjunction with another source of energy. Flywheels have the advantage of greater specific power than batteries and are a very suitable method of storing energy from regenerative braking. This approach has been successfully applied to racing cars, such as the Le Mans Prototype 1 (LMP1) Audi R18 e-tron quattro shown in Figure 1.32.

Figure 1.32: Audi R18 e-tron quattro uses regenerative braking to charge a flywheel to provide additional power during acceleration. Image from Max Earey, Shutterstock.com; https://www.shutterstock.com/image-photo/goodwood-uk-july-1-audi-r18-107619215?src=c2KZUbELgBNwk95PiyTiAg-1-4.

Transit System Energy Recovery

Flywheels may be used to recover lost energy from electric trains used for public transit. A flywheel energy storage system may be installed at a transit station and energy produced during regenerative braking when the train stops at the station can be stored in the flywheel. This energy may then be used to provide power to the train for acceleration when it leaves the station. This approach recovers energy that would otherwise be lost and helps to reduce voltage sag on the transit power line, which can occur when additional load is placed on the line by accelerating trains. Demonstration systems have been installed at transit stations in Los Angeles, New York, London, Lyon, and Paris and have provided positive evidence for the utility of this approach.

Spacecraft Energy

NASA has been involved in the development of a flywheel energy storage system for possible spacecraft applications. Flywheels have potential advantages over traditionally used battery storage systems. These include

- high power density,

- long life,

- deep depth of discharge, and

- broad operating temperature range.

A diagram of NASA's prototype unit (named G2) is shown in Figure 1.33. The flywheel consists of a carbon fibre rim attached to a titanium hub and rotates at a maximum of 60,000 rpm. This unit stores a total of 525 Wh of energy and has an output of 1 kW, for a duration of just over half an hour. A photograph of the prototype unit is shown in Figure 1.34.

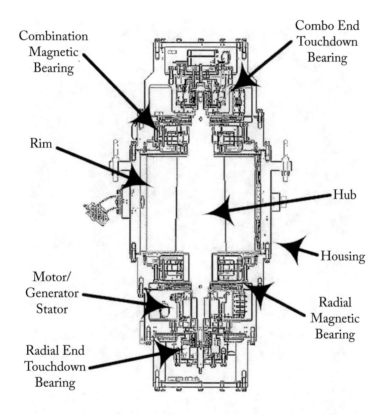

Figure 1.33: **Diagram of NASA's G2 flywheel energy storage module;** https://ntrs.nasa.gov/archive/nasa/casi.ntrs.nasa.gov/20060028492.pdf.

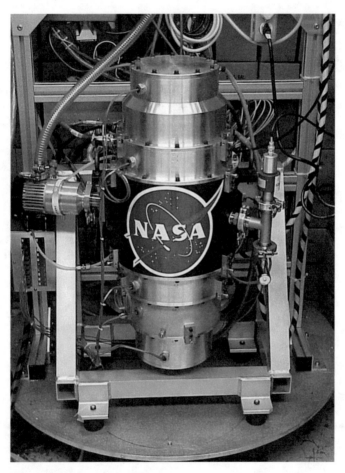

Figure 1.34: Photograph of NASA's G2 flywheel energy storage module; https://commons.wikimedia.org/wiki/File:G2_front2.jpg.

Nuclear Fusion Experiments

Many nuclear fusion experiments use very large amounts of power for relatively short periods of time. Flywheels are an ideal means of storing electrical energy for these short power bursts. The Joint European Torus (JET) is an example of a facility that uses flywheel energy storage for fusion research. The JET is the world's largest magnetic confinement fusion reactor. It is a tokamak-based reactor design and is located at the Culham Centre for Fusion Energy in the United Kingdom. Typical experimental runs (or "shots") last about 20 sec and can use up to 1 GW of peak electrical power. Since this is more than can be supplied by the electric grid, energy is stored in two flywheels that are used to provide electricity during the experiment. Each flywheel has a 9-m diameter rotor and weighs over 700 tonnes. At full speed the rotors turn at 225 rpm and they are cycled down to about half speed during the 20 sec discharge. Each flywheel can provide up to about 500 MW output. Recharging the flywheels takes about 9 min and is accomplished by two 8.8 MW (~12,000 hp) motors, one for each rotor. This represents an ideal application of flywheel energy storage where high power and reasonably rapid cycle time is needed.

CHAPTER 2

Thermal Energy Storage Methods

2.1 INTRODUCTION

Energy may be stored by using the thermal properties of a material. These techniques can be utilized on a wide variety of scales, from individual, residential energy storage to storage on a city or regional scale. Three different material properties allow for the storage of thermal energy:

1. sensible heat,

2. latent heat, and

3. thermochemical reactions.

Sensible heat is related to the change in temperature of a material to which heat is applied that results from its heat capacity. Energy storage that uses the sensible heat capacity of a material is described for some commonly used materials in Section 2.2. Latent heat is the heat that is related to a phase transition of a material. This may be (for example) the heat of fusion if the material undergoes a transition between a solid and a liquid phase or the heat of vaporization if the material undergoes a transition between a liquid and gas phase. Energy storage related to the latent heat of a material is described in Section 2.3. Thermochemical energy storage methods involve chemical reactions such as dehydration of metal hydrides or salt hydrates. These methods are discussed in Section 2.4.

2.2 SENSIBLE HEAT ENERGY STORAGE

Basics of Sensible Heat Energy Storage

The sensible heat associated with a material is the energy that is needed to heat the material from a lower temperature to a higher temperature when there are no phase transitions between the two temperatures. It is also the energy that can be extracted from the material by cooling it from the higher temperature to the lower temperature. The relationship between the thermal energy, Q, and the change in temperature, ΔT, is

$$Q = C\Delta T, \qquad (2.1)$$

where C is the heat capacity of the material. When Q is in Joules and temperature is in Celsius (or Kelvin) then the heat capacity is in units of J/°C. When selecting materials for a particular application it is generally reasonable to consider the intrinsic properties of the material, that is the heat capacity for a particular quantity of the material. This may be expressed as the heat capacity per unit mass, c (called the *specific heat capacity*, or more commonly, just the *specific heat*), the heat capacity per mole, C_{mol} (called the molar heat capacity), or the heat capacity per unit volume, C_{vol} (called the volumetric heat capacity). The relationship of these quantities to the heat capacity is given by

$$c = C/m, \qquad (2.2)$$

where m is the sample mass,

$$C_{mol} = CM/m, \qquad (2.3)$$

where M is the molecular weight of the material and

$$Cvol = C/V = c/\rho, \qquad (2.4)$$

where V is the sample volume and ρ is the material density. The volumetric heat capacity is generally only used for liquids and solids and not gases.

2.2.1   SENSIBLE HEAT ENERGY STORAGE MATERIALS

For particular heat storage applications, the choice of a storage material must consider the thermal properties of that material. Some relevant properties are:

- the phase of the material, i.e., solid or liquid, may be of relevance to the way in which the stored heat will be distributed and used;

- the volumetric heat capacity is of relevance if the space for the storage facility is limited;

- the specific heat is important if the mass of the storage facility is limited; and

- the range of temperatures over which the material can be used (i.e., temperatures for which it is stable and does not undergo any phase transitions).

Other factors such as toxicity and cost should also be considered. Table 2.1 gives the relevant properties of some common materials that are used for thermal energy storage. Clearly, from an energy standpoint, water stands out as a suitable material for heat storage.

Table 2.1: Typical values for thermal properties of some common materials for sensible heat energy storage

Material	Specific Heat [J/(kg · °C)]	Density [kg/m³]	Volumetric Heat Capacity [kg/(m³ · °C)]
Water	4,186	1,000	4,186
Wood (pine)	2,800	500	1,400
Stone (solid)	879	2,560	2,250
Stone (loose)	879	1,500	1,320
Sand	816	1,600	1,306
Concrete	653	2,300	1,502
Iron	460	7,855	3,613

2.2.2   RESIDENTIAL HEAT STORAGE

Perhaps the simplest and most obvious application for sensible heat energy storage is for residential use. Residences which utilize solar panels for space heating or domestic hot water heating purposes require diurnal heat storage to provide thermal energy during the night or during periods of low solar insolation. Water is an appropriate storage material, not only from a thermal properties standpoint, but because it is a common non-toxic liquid that is suitable for distributing heat through a central heating system or for providing domestic hot water.

There are several different approaches to the use of stored energy from solar collectors. Figure 2.1 shows a simple passive system for heating water for domestic hot water use. Figure 2.2 shows a photograph of this type of installation. The storage tank is mounted on the roof immediately adjacent to the solar collector. Pressurized cold water runs through the tank and is heated and returned as hot water for residential use. The system shown in Figure 2.1 uses an auxiliary natural gas-fired heater to maintain appropriate water temperature. Such systems may not be suitable for colder climates because of heat loss from the external storage tank.

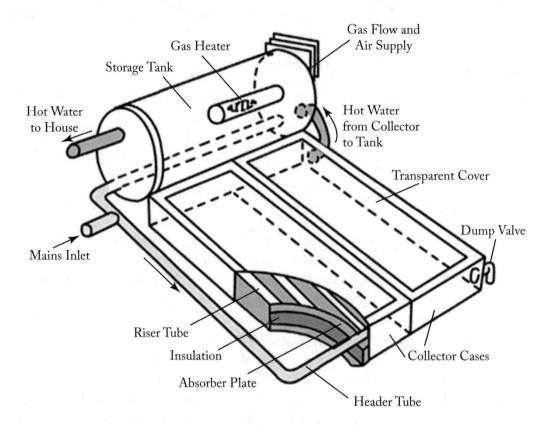

Figure 2.1: Diagram of a solar thermal domestic hot water system using a flat plate collector and a roof-mounted storage tank. Based on http://www.yourhome.gov.au/energy/hot-water-service.

Figure 2.2: A roof-mounted solar storage tank coupled to a flat plate collector. Image from Shutterstock.com; https://www.shutterstock.com/image-photo/solar-panel-hot-water-system-on-188171915?src=CX5vquGElCSpc-IvygOigw-2-91.

An active solar heating system, i.e., one in which water or a working fluid is pumped through a solar collector and storage tank is illustrated in Figure 2.3. In this case, a ground level or basement storage tank can be used, as shown in the figure. This arrangement has the potential of utilizing an internal storage tank and thereby reducing the heat loss that is characteristic of an external tank. Figure 2.4 shows a schematic of a solar space heating system using a heat exchanger in the storage tank. Figure 2.5 shows a photograph of a typical basement installation of a solar-heated water storage tank for space heating purposes.

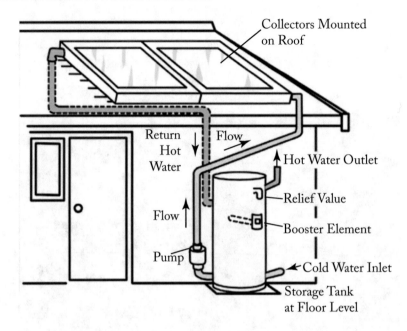

Figure 2.3: Active solar heating system utilizing roof mounted flat plate collectors and a ground level storage tank. Based on http://www.yourhome.gov.au/energy/hot-water-service.

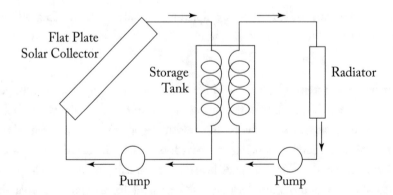

Figure 2.4: Simple system for storing thermal energy from a flat plate solar collector for residential use during the night.

Figure 2.5: Photograph of interior solar thermal storage tank. Image from Shutterstock.com; https://www.shutterstock.com/image-photo/installing-solar-water-tank-boiler-room-1035247072?src=CX5vquGElCSpc-IvygOigw-2-2.

The design of a heat storage system, as illustrated in Figures 2.3, 2.4, and 2.5, for space heating purposes requires careful consideration of the area of the solar collectors to ensure sufficient energy to satisfy heating requirements. It is also necessary to consider the size of the storage tank in order to provide adequate heat during the night and during days of low insolation. These factors obviously relate to the local climate as well as details of the building construction.

As an example, we can estimate the winter diurnal heating requirement for a typical house with a volume of 700 m^3. A reasonably well-insulated house might require 2.8 kJ/m^3 per hour per °C temperature difference between the inside of the house and the outside. We consider an average daily outside temperature of -6°C (typical of a cold winter day in Boston, MA) and require an inside temperature of 19°C. A simple calculation gives the daily heating requirement as

$$(2800 \text{ J/m}^3 \text{ per °C per hour}) \times (24 \text{ h/d}) \times (700 \text{ m}^3) \times (25°C) = 1.18 \times 10^9 \text{ J/d} . \qquad (2.5)$$

To determine the size of a suitable storage system as shown in Figure 2.4, we combine Equations (2.1) and (2.2) and solve for mass:

$$m = \frac{Q}{c \Delta T} . \tag{2.6}$$

The temperature difference, ΔT, in this equation gives the difference in the temperature of the storage tank when it is at a maximum after being heated from the solar collector, T_{max}, and when it is at a minimum after being used to heat the house overnight, T_{min}. Using an estimate of these temperatures as $T_{max} = 85°C$ and $T_{min} = 35°C$, we find the mass of water needed to store the energy given in Equation (2.6) as

$$m = \frac{1.18 \times 10^9 \text{J}}{4,186 \dfrac{\text{J}}{\text{kg} \times °C} \times (85°C - 35°C)} = 5,600 \text{ kg}. \tag{2.7}$$

This mass represents a volume of water of 5.6 m^3. This can be stored in a cylindrical tank 2 m high by about 2 m in diameter.

On the basis of Table 2.1, it is possible to consider materials other than water for residential heat storage. From the materials in the table, iron is seen to be a close second to water in terms of volumetric heat capacity. Although one does not think of iron as a particularly expensive material, a calculation of the required mass of iron to replace water for the calculation in Equation (2.7) yields a value of around 50,000 kg, which gives an unreasonable cost, even for a modestly priced material. It is seen from the table that stone may be the most reasonable alternative to water. The specific heat is roughly twice that of iron and it is, as a naturally occurring material, readily and inexpensively available. A typical system utilizing loose stone for residential heat storage is illustrated in Figure 2.6. Although the volumetric heat capacity is only a little more than half that of water, such a system may have advantages in some situations. As the figure shows, loose-packed material, specifically rock, can be used efficiently as a heat storage medium for hot air space heating systems. This provides a simpler alternative to water-based heat storage for hot air heating systems, where a water-to-air heat exchanger is needed. The use of rocks for heat storage, combined with air as the working fluid to transport heat, does not limit the operating temperature to less than 100°C, as is the case for water-based systems. While this is not generally a consideration for a residential system incorporating flat-plate solar collectors, it is a factor for concentrating solar collector systems that are commonly used for grid-integrated systems as discussed below. A number of studies have been undertaken to evaluate parameters such as rock size and packing fraction, along with air flow rate, on system performance (e.g., Choudhury et al., 1995, Singh et al., 2013).

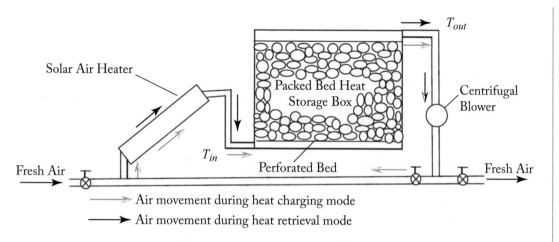

Figure 2.6: Residential thermal storage using the heat capacity of rocks. Based on Singh et al., 2015, with permission from Elsevier.

2.2.3   COMMUNITY-BASED HEAT STORAGE

While the system described above is suitable for an individual residence, the use of thermal storage for an entire community, i.e., housing development, town, city, etc. requires a system with a much greater heat storage capacity. It would also require a greater capacity to harvest heat to be stored. In this section, three community-based systems for heat storage are described: the district heating system of Reykjavik, Iceland; the heat storage system for the Drake Landing Solar Community in Okotoks, Alberta, Canada; and the heating system in Brædstrup, Denmark.

Reykjavik District Heating System

Geothermal energy can be utilized as a source of electricity as well as a source of thermal energy. Direct geothermal energy is utilized by a number of major cities worldwide who fulfill a portion of their heating needs from this source. These cities include Beijing, Bucharest, Budapest, Paris, Rome, and Sophia.

Iceland has abundant geothermal resources. These resources provide about 30% of the electricity used in the country and nearly 90% of the country's 338,000 residents obtain space heating from geothermal resources. There are 29 separate district heating systems in Iceland that provide heat to various communities. In the capital city, Reykjavik, 95% of the buildings are connected to a centralized district heating system which utilizes geothermal energy.

Figure 2.7 shows a simplified diagram of the system that provides geothermal heating to Reykjavik. Geothermal water is provided by four geothermal fields. One of these fields, Nesjavellir, is a high-temperature geothermal resource while the other three fields are low-temperature resources. The largest of the three low-temperature resources is Reykir in Mosfellsbær, which provides 1,700 kg/s of water at 85–90°C. Storage of geothermal heat is an important component of the Reykjavik district heating system. The use of geothermal water from Reykir is an example of how storage facilities are important for the distribution of heat to the community. The geothermal water from Reykir flows to a storage facility consisting of six tanks outside of Reykjavik which have a total capacity of 5.4×107 L. From there, the water flows to the city where it fills six additional storage tanks with a capacity of 2.4×107 L. This latter facility, known as the Perlan (Icelandic for "The Pearl"), sits atop Öskjuhlíð, a hill in central Reykjavik, which is 61 m above sea level. Figure 2.8 shows a photograph of the Perlan. The central dome above the storage tanks houses a rotating restaurant, a planetarium, and a gift shop. From the Perlan, geothermal water is distributed to the city through nine pumping stations. Because district heating requirements are not constant in time, the storage tanks provide a buffer for periods of high demand.

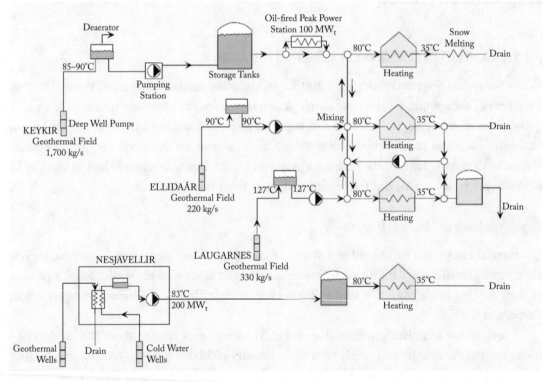

Figure 2.7: Simplified diagram of the district heating system in Reykjavik, Iceland. Based on Gunnlaugsson, 2003, reproduced with permission.

Figure 2.8: The Perlan geothermal storage tanks in Reykjavik, Iceland. Image courtesy of Richard A. Dunlap.

Drake Landing Solar Community

The Drake Landing Solar Community in Okotoks, Alberta, Canada is a community consisting of 52 single-family homes, as shown in Figure 2.9. The community was constructed in 2005 and the homes range in size from 139–155 m^2 in floor area. The community is designed to obtain nearly all its heating from solar energy. The goal of the design is to supply virtually all the heat required during the winter months, when the heating requirements are the greatest and the insolation is the least, using thermal energy that is stored during the summer months when heating requirements are minimal. Domestic hot water is also supplemented with solar energy.

There are 800 single-glazed flat-plate solar collectors mounted on 45-m^2 detached garages as seen in the figure. The garages are interconnected to one another with breezeways in order to provide a contiguous platform for mounting the solar collectors. In the summer, these collectors provide an average of about 1.5 MW of thermal power during the day. The basic design of the community heating system is illustrated in Figure 2.10. Fluid (water with non-toxic glycol added) is circulated through the solar collectors, where it is heated, and then into a series of 144 boreholes, each about 35-m deep, where the heat is transferred to the soil. Details of the borehole design are shown in Figure 2.11. Electricity for the pumps required to circulate the fluid is obtained from photovoltaic collectors to ensure net carbon-neutrality for the system.

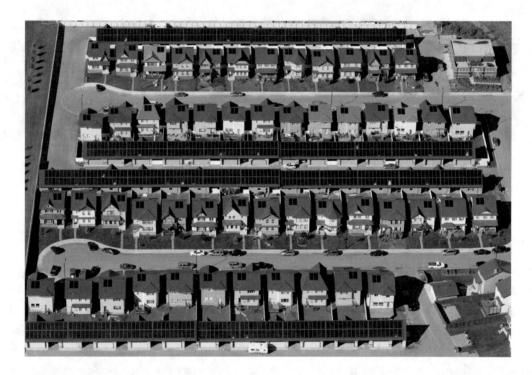

Figure 2.9: Aerial photograph of the Drake Landing Solar Community in summer; https://www. nrcan.gc.ca/energy/publications/sciences-technology/buildings/17864, reproduced with permission from Natural Resources Canada.

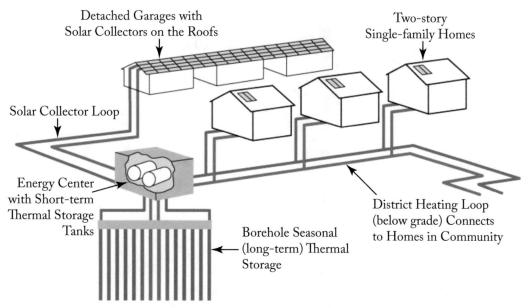

Figure 2.10: Schematic diagram of the design of the heating system for the Drake Landing Solar Community in Alberta, Canada. Based on https://www.dlsc.ca/how.htm.

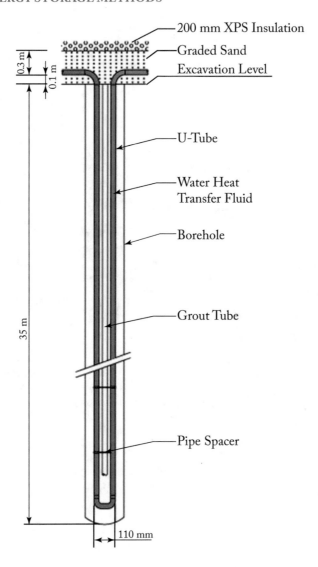

200 mm XPS Insulation

Graded Sand

Excavation Level

U-Tube

Water Heat
Transfer Fluid

Borehole

Grout Tube

Pipe Spacer

0.3 m

0.1 m

35 m

110 mm

Figure 2.11: Details of the borehole construction at the Drake Landing Solar Community. Based on https://www.dlsc.ca/borehole.htm.

By the end of the summer, the soil in the vicinity of the boreholes reaches a temperature of about 80°C. When heat is required in the homes, cold water (with glycol) is circulated through the boreholes where it is heated. It is then circulated through heat exchangers in the homes to provide space heating as well as domestic hot water. Since the community was established, more than 90% of the required heat has been provided by solar energy, as well as about 50% of the domestic hot water. The remaining heat and hot water are provided by the combustion of natural gas.

Brædstrup Fjernvarme

Brædstrup is a town in Denmark (population about 3,600) that has taken an approach to seasonal thermal energy storage that is similar to that of the Drake Landing Solar Community in Canada. The Brædstrup Fjernvarme (Fjernvarme = Danish for heating) system uses flat-plate solar collectors to heat water, which is then used to heat soil through bore holes. The system was developed in two stages. Stage I, which utilized 8,000 m^2 of solar collectors, became operational in 2007. Stage II, which expanded the solar collector area to 18,600 m^2, became operational in 2011.

Figure 2.12: Solar collectors at Brædstrup Fjernvarme energy facility; https://www.skyfish.com/p/danskfjernvarme/425141/12317286?predicate=created&direction=desc.

The Brædstrup Fjernvarme system is a hybrid solar-natural gas system. The solar collectors provide thermal energy while the natural gas generating station is a combined heat and power (CHP) facility that produces both electricity and heat. The electricity produced by the natural gas generating station is fed into the grid. The heat produced from the natural-gas-fired station is combined with heat from the solar collectors and is stored in two above-ground thermal water storage tanks. Long-term (i.e., seasonal) heat storage is accomplished, as for the Drake Landing Community, by heating soil by circulating hot water through bore holes. In the winter, heat stored during the summer is recovered by circulating cold water through the bore holes.

At present, the solar-natural gas heating system as described above provides 95% of the heat required by the town of Brædstrup. The solar contribution to the total heat supplied is about 20%, while the remaining 80% comes from the thermal output of the natural gas CHP plant.

Figure 2.13: Interior of the Brædstrup Fjernvarme energy facility; https://www.skyfish.com/p/dansk-fjernvarme/425141/12580016?predicate=created&direction=desc.

2.2.4   SOLAR PONDS

A solar pond is an artificial pond that is analogous to a flat-plate solar collector coupled to a thermal water storage tank. That is, it converts solar radiation into heat and stores the resulting thermal energy. The development of solar ponds follows from the observation that some natural salt lakes develop large thermal and salinity gradients during the summer months. In some cases, surface water may be nearly fresh at a temperature of around 25°C, while water at a depth of 1–2 m may be a nearly saturated salt solution at a temperature of 50–60°C. This phenomenon may be applied to the construction of an artificial solar pond where energy is acquired from solar radiation and is stored as a result of the natural convective structure of the pond.

Figure 2.14 shows the relevant features of a simple salinity gradient solar pond. The pond is divided vertically into three distinct zones. The zone at the top of the pond is the upper convective zone (UCZ) where the temperature and salinity are approximately constant. The middle zone is the non-convective zone (NCZ) where there are both temperature and salinity gradients. As seen in the figure, both temperature and salinity increase with increasing depth below the surface in the non-convective zone. The bottom zone is the lower convective zone (LCZ) where, again, the temperature and salinity are more or less constant. Typically, the lower convective zone is a concentrated brine solution. Heat is stored (and extracted) from the lower convective zone. The non-convective zone acts to insulate the warmer lower convective zone from the cooler upper convective zone.

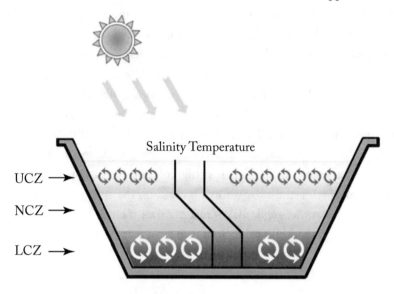

Figure 2.14: Structure of a solar pond showing the upper convective zone (UCZ), non-convective zone (NCZ), and lower convective zone (LCZ). The salinity and temperature profiles are shown for the three zones. Based on Khalilian, 2017, with permission from Elsevier.

Solar radiation which is incident upon the pond interacts first with the upper convective zone. Some of the radiation will be reflected from the surface and will not contribute to the heat content of the pond. Some of the incident solar radiation will be absorbed as heat in the upper convective zone. Most of this is lost back to the atmosphere through radiation or convection. A fraction of the incident solar radiation will be transmitted through the upper layer and will be absorbed in the non-convective zone. Some will also pass through the non-convective zone and be absorbed as heat in the lower convective zone. Heat which is absorbed in the lower convective zone is trapped because of the insulating non-convective zone above it. The resulting temperature difference between the upper and lower convective zones can be as large as 60°C. It is the heat that is trapped in the lower convective zone that can most readily be recovered. Thermal energy from solar ponds has been used for space heating, heat for industrial processes, heat for swimming pools and electricity generation. Recent studies (Leblanc et al., 2011) have shown that overall thermal efficiency may be improved by also utilizing the heat content in the non-convective zone.

Figure 2.15 shows a simple method of recovering thermal energy from the lower convective zone of a solar pond. Hot concentrated brine is pumped from the lower convective layer through an external heat exchanger where the heat content is transferred to water flowing through the heat exchanger. The potential difficulty of this approach is that the brine which is returned to the lower convection zone after passing through the external heat exchanger is colder, and therefore denser, than the hot brine already in this zone, and tends to lie on the bottom without mixing efficiently with the warmer brine in this zone. However, this method has been successfully used for the solar pond located in Beit HaArava, an Israeli settlement in the West Bank (see below).

Another method of extracting heat from the solar pond is through the use of an internal heat exchanger as illustrated in Figure 2.16. Here cold fresh water is circulated through a heat exchanger located in the lower convective zone. The fresh water is heated and can then be used (for example) for space heating, either as hot water or the heat can be transferred to air with an external water-to-air heat exchanger, as shown in the figure. While this alleviates the problem related to mixing associated with the external heat exchanger in Figure 2.15, the system is more complex, and the location of the internal heat exchanger makes maintenance more involved. This method has been successfully used at the Pyramid Hill solar pond facility in Australia (see below).

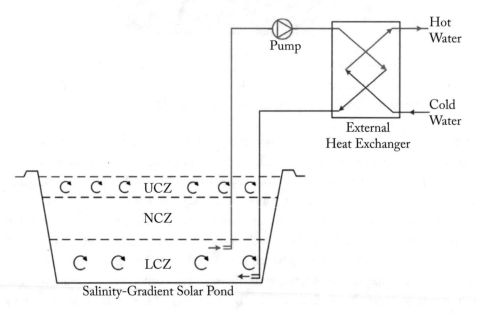

Figure 2.15: Heat recovery from the lower convective zone of a solar pond using an external heat exchanger. Based on Leblanc et al., 2011, with permission from Elsevier.

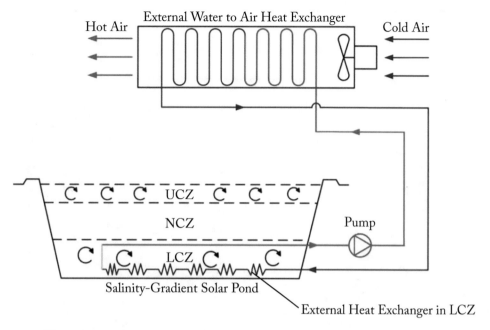

Figure 2.16: Heat recovery from a solar pond using an internal heat exchanger located in the lower convective zone. Based on Leblanc et al., 2011, with permission from Elsevier.

In addition to providing heat for potential industrial processes or space heating purposes, thermal energy from solar ponds can be used to generate electricity. A simple design for a system to produce electricity from a solar pond is shown in Figure 2.17. As the hot brine from the lower convection zone is typically at a temperature of 80–90°C, an organic Rankine cycle turbine coupled to a generator is used to produce electricity. The heat from the hot brine is used to evaporate a working organic fluid with a low boiling temperature and this vaporized organic fluid is used to drive the turbine. Typical organic fluids that are suitable for this application include freon and propane.

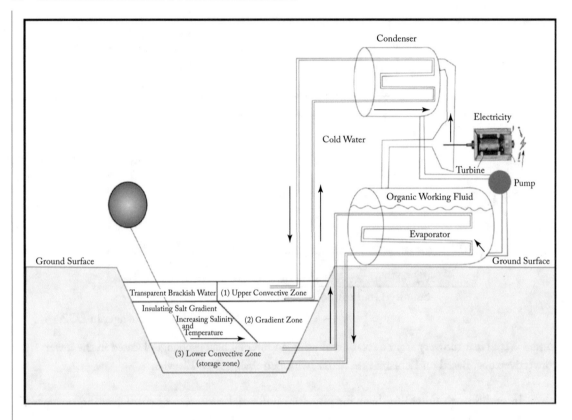

Figure 2.17: Method of using heat from a solar pond for the generation of electricity by means of an organic Rankine cycle turbine. Based on Howari et al., 2008.

While solar ponds have not been commercialized extensively, there are some notable examples of the use of this technology. The largest solar pond constructed to date was built in Beit HaArava, an Israeli settlement in the West Bank in 1983. It operated until 1988. This 210,000 m^2 pond was used for electricity generation and produced 5 MW electrical output.

A solar pond constructed in 2001 in Pyramid Hill, Australia is shown in Figure 2.18. The thermal energy from this 3,000 m^2 pond is used for commercial salt production and aquaculture. The plastic rings seen on the surface of the pond are designed to minimize adverse effects from wind-driven waves.

Figure 2.18: The 3,000 m^2 solar pond at Pyramid Hill, Australia. Reprinted with permission from Elsevier (Malik et al., 2011).

The El Paso solar pond, shown in Figure 2.19, was a 3,700 m^2 pond constructed by the University of Texas in 1983 which operated until 2003. It was located on the property of the canning company Bruce Foods, Inc. The pond provided a convenient research platform for testing solar pond technologies. It also provided thermal energy for industrial operations at Bruce Foods and an organic Rankine turbine generated an electrical output of 100 kW that was fed into the local grid.

Figure 2.19: **Solar pond in El Paso, Texas** (Lu, Walton, and Hein, 2002, https://www.usbr.gov/research/dwpr/reportpdfs/report080.pdf).

2.2.5   GRID INTEGRATED SYSTEMS

The most notable examples of grid-integrated generating facilities that utilize thermal energy storage are probably concentrating solar collectors. These typically use the sensible heat associated with molten salt to store thermal energy. Figure 2.20 shows a typical concentrating central receiver system. The same concept can also be applied to parabolic receiver systems. "Cold" molten salt is stored in the cold salt storage tank. It is pumped through the focus area of the central receiver where it is heated and is then pumped to the hot salt storage tank. To generate electricity, the hot molten salt is pumped through a heat exchanger to heat water and to produce steam. This steam is then used to drive a turbine/generator to produce electricity. The molten salt is then returned to the cold storage tank. Salt storage tanks at a typical parabolic trough solar-generating station are shown in Figure 2.21.

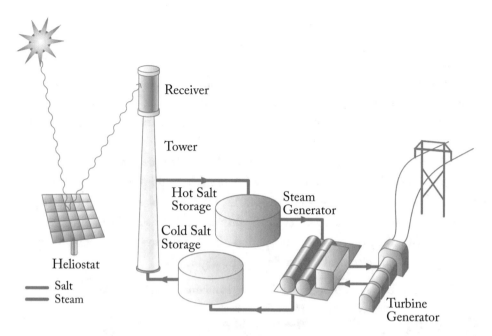

Figure 2.20: Diagram of the Solar Two facility in the Mojave Desert in California showing thermal energy storage in molten salt tanks. Based on https://www.nrel.gov/docs/legosti/fy97/22835.pdf.

Figure 2.21: Molten salt storage tanks at the Solana Generating Station a parabolic trough solar facility near Gila Bend, Arizona; https://commons.wikimedia.org/wiki/File:Abengoa_Solar_(7336087392).jpg.

The salt that is most commonly used for systems as shown in Figure 2.21, is a eutectic mixture of 60% KNO_3 and 40% $NaNO_3$. Some relevant properties of this salt mixture are given in Table 2.2. The cold salt storage tank stores the molten salt at a temperature of around 285°C and the hot salt storage tank stores the heated salt at a temperature of around 565°C. Well-insulated storage tanks can maintain the salt temperature for about a week and the estimated efficiency of thermal storage is around 99%, meaning that the electricity generated after thermal storage of the salt is 99% of that which would have been generated immediately. The size of the storage tanks depends on the required power output and duration. As an example, two tanks (one cold and one hot) with dimensions of 24.4 m in diameter by 9.1 m high, could provide an electrical output of 50 MW for a period of 8 hr. In the summer of 2013, the Gemasolar Thermosolar central receiver power plant in Fuentes de Andalucía, Seville, Spain was the first such facility to generate power continuously for 24 hr/day over an extended period of time (36 days) using thermal energy stored during the day to provide electrical output during the night.

Table 2.2: Some relevant properties of a eutectic mixture of 60% KNO3 and 40% NaNO3 salt compared with water and loose rock

Property	Salt	Water	Rock	Units
Melting point	240	0	–	°C
Density	2200	1000	1500	kg/m3
Specific heat	1530	4186	879	J/(kg·°C)
Volumetric heat capacity	3366	4186	1320	kJ/(m3·°C)

It has also been proposed that rock beds could be used for thermal storage for concentrating solar collector systems. A possible system for this purpose is illustrated in Figure 2.22. Air is heated by concentrated solar radiation in the central receiver. This hot air is then circulated thorough a bed of loose rock. The hot rock stores the thermal energy and this energy is recovered as needed and transferred to water in a heat exchanger to produce steam to run turbines. A comparison of the properties of salt, water, and rock is shown in Table 2.2. While rock has a lower heat capacity than water or salt, it is useable over a greater range of temperatures.

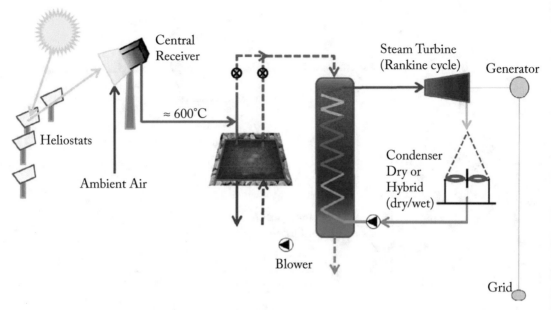

Figure 2.22: Schematic of concentrating solar system utilizing a rock bed for thermal storage. Based on Allen et al., 2014, with permission from Elsevier.

2.3   LATENT HEAT ENERGY STORAGE

Latent heat energy storage takes advantage of the large amount of heat that accompanies phase changes in a material. Typical examples of phase transitions are the transitions between the solid, liquid, and gaseous forms of a material. The energy associated with the phase transition between the solid state and the liquid state (melting) is the latent heat of fusion and the energy associated with the phase transition between the liquid state and the gaseous state is the latent heat of vaporization. Figure 2.23 shows the temperature of a material a function of the energy input and illustrates the latent heat associated with the phase transitions.

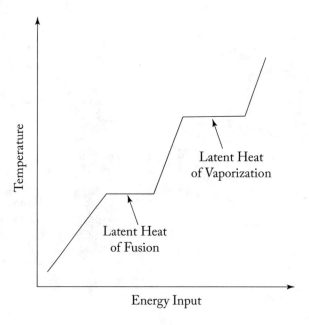

Figure 2.23: Temperature as a function of energy input for a material showing the latent heat of fusion and the latent heat of vaporization.

Table 2.3 shows a comparison of the latent heats associated with the phase transitions in water and a comparison with the sensible heat associated with raising the temperature of water from 0°C to room temperature (20°C) and from room temperature to 100°C. It is clear from the table that the latent heat associated with phase transitions can be much larger than the sensible heat associated with temperature changes of the material. This is particularly the case for the latent heat associated with the liquid to gas phase transition.

Table 2.3: A comparison of the latent heats and the sensible heat of water. Quantities are per kg of water

Heat	Value (kJ/kg)
Latent heat of fusion	334
Sensible heat 0–20°C	84
Sensible heat 20–100°C	335
Latent heat of vaporization	2260

The large change in the heat content of a material associated with a phase transition at a temperature, T, may be explained in terms of the Gibbs free energy, G, which is defined as

$$G = H - TS.$$

(2.8)

Here H is the heat content of the material and S is the entropy. At a phase transition, the change in the Gibbs free energy is

$$\Delta G = \Delta H - T\Delta S, \tag{2.9}$$

where, as shown in Figure 2.23, the temperature remains constant. As the Gibbs free energy at a phase transition is a continuous function of temperature, the change in Gibbs free energy must be zero. This means that

$$\Delta H = T\Delta S, \tag{2.10}$$

and this shows that, as the entropy (degree of disorder) changes when a material undergoes a change from solid to liquid or from liquid to gas, there is a change in its heat content.

The above example of phase changes in water is only one of the possible approaches to using latent heat for energy storage. The latent heat of fusion associated with the melting transition of other materials may also be useful for the purpose of energy storage and some examples of potential materials are shown in Table 2.4. In addition, the latent heat associated with solid-solid phase transitions (i.e., crystallographic phase transitions) is also a possible means of storing energy, although the heat associated with such transitions is typically much smaller than the latent heat of fusion or vaporization. Three examples of the use of latent heat of materials for energy storage are presented in the forthcoming sections.

Table 2.4: Melting temperatures and heats of fusion for some materials

Material	Melting Temperature (°C)	Density (kg/m³)	Heat of Fusion (kJ/kg)
Polyethylene glycol 400	8	1130	100
Caprylic acid	16	910	149
Decanoic acid	32	893	153
Dodecanoic acid	43	930	184
Paraffin	64	900	174
Stearic acid	69	941	203
NaOH	318	2130	150
FeCl$_2$	670	3160	340
KCl	776	1980	340
NaCl	801	2150	500

2.3.1   RESIDENTIAL LATENT HEAT ENERGY STORAGE

In Section 2.2, we saw that the sensible heat of water was a convenient means of storing heat from a solar collector for residential use during the night. The use of materials which undergo a phase transition (e.g., melting) within the temperature range of the operation of the storage facility have certain advantages over the use of sensible heat storage using water. The latent heat adds to the sensible heat storage and by properly matching the thermal properties of the storage material with the operating temperature range of the storage system and the heating requirements, some potential advantages may be seen. These include:

- use of a smaller mass and/or volume of thermal storage material and

- use of a smaller range of operating temperatures.

These points are illustrated in Figure 2.24 by a comparison of the heat storage of dodecanoic acid (labeled PCM = phase change material) and water over a temperature range that includes the melting temperature of the dodecanoic acid. It is seen from an investigation of the figure that over a temperature range of (say) 35–50°C, 60 kg of dodecanoic acid provides about 12 kJ of available heat, compared to about 3 kJ for the same mass of water.

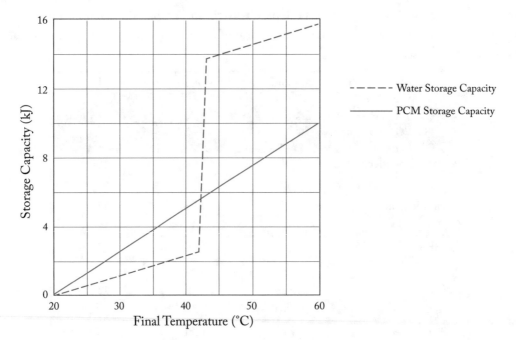

Figure 2.24: Comparison of storage capacity of 60 kg of dodecanoic acid (PCM) and 60 kg of water. Based on Kabbara et al., 2016, reproduced with permission.

2.3.2   LATENT HEAT ENERGY STORAGE IN ICE

The use of the latent heat of fusion of water is a convenient means of load leveling in order to provide air conditioning. In warm summer weather, the air conditioning of large buildings represents one of the most significant contributions to electricity use in many cities and load leveling using the latent heat of water can significantly reduce these requirements. Ice is produced by refrigeration techniques in the night when cooling needs are minimal. At this time, electricity demand is minimal and, in addition, electricity rates are low. The ice is then used to cool air to provide air conditioning during the day when electricity rates are high and the need for air conditioning is the greatest. This approach has two benefits: it provides a more economic means of air conditioning a building by utilizing electricity at low rate times and it reduces the need for additional grid generating capacity to cover periods of high demand.

This technology is probably best suited for large office buildings and similar structures. Many major buildings worldwide utilize this technology, including the world's tallest building, Burj Khalifa in Dubai, United Arab Emirates; see Figure 2.25.

2.3.3   LATENT HEAT ENERGY STORAGE IN LIQUID AIR

Another approach to the storage of energy is the use of the latent heat of vaporization of air. In this scheme, excess electricity that is available during the low rate period is used to liquefy air according to the diagram shown in Figure 2.26. The air is first compressed, as discussed in Section 1.5, for compressed air energy storage. This gives rise to the generation of heat as previously described. This heat is stored for use during the expansion phase of energy recovery. The compressed air is then liquefied and stored.

Figure 2.25: Burj Khalifa in Dubai, United Arab Emirates. This is the world's tallest building (as of 2019). Image from Melinda Nagy, Shutterstock.com; https://www.shutterstock.com/image-photo/dubai-uae-february-2018-burj-khalifa-1032298924?src=ac0Z0oKrZoW49vWb3yW3JQ-1-44.

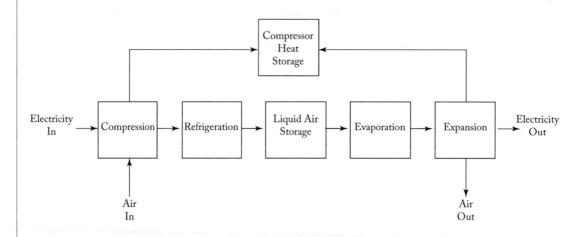

Figure 2.26: Schematic of liquid air energy storage using the latent heat of vaporization.

When additional electricity is required during periods of high demand, the energy stored in the liquid air is recovered, as shown in Figure 2.25. Following along the lines of the energy recovery from compressed air energy storage, as described in Section 1.5, the liquid is evaporated and expanded to operate a turbine/generator to produce electricity. During the expansion phase, heat from the compression phase is recovered.

Liquid air energy storage is a potential technology for grid storage to cover periods of peak demand and is, therefore, a possible alternative to pumped hydroelectric storage or traditional compressed air energy storage.

Figure 2.27: Experimental 350 kW liquid air energy storage facility in Slough, United Kingdom; https://www.birmingham.ac.uk/research/energy/news/2018/university-of-birmingham-research-applied-to-build-worlds-first-liquid-air-energy-storage-plant.aspx.

2.4   THERMOCHEMICAL REACTIONS

Thermal energy can be stored in materials that undergo reversible chemical reactions. The basic principle follows from the fact that the reaction will be endothermic in one direction and exothermic in the opposite direction. Heat may be stored by causing the endothermic reaction to occur and that heat may be recovered by causing the exothermic reaction to occur. A number of different reactions have been considered for thermochemical energy storage. Although the application of this approach is still in the early stages of development, several types of materials, such as salt hydrates, have shown some promise.

The reversible reaction involving the dehydration and hydration of salt hydrates is of the form

$$\text{salt hydrate} \leftrightarrow \text{salt} + \text{water}. \tag{2.11}$$

A typical example of this type of reaction that has been investigated as a possible means of thermochemical energy storage is

$$MgSO_4 \cdot 7H2O \leftrightarrow MgSO_4 \cdot H_2O + 6H_2O. \tag{2.12}$$

Thermal energy is stored during the dehydration of the salt hydrate by heating. Energy is recovered during the hydration of the salt by exposing it to water or water vapor.

The dehydration of salt hydrates as a means of thermal energy storage has a number of attractive features.

- The materials are common and relatively inexpensive.

- The chemical needed for hydration and energy recovery (i.e., water) is universally available.

- After dehydration the salt may be stored in a sealed container indefinitely without loss of energy.

- Theoretical energy storage capacity per unit mass can be as much as 10 times that of sensible heat energy storage in water.

- After dehydration the salt may be transported to another location (if appropriate) for hydration (and energy release).

These materials present an attractive alternative to sensible heat storage in water for residential heat storage because of the much smaller mass and volume of material needed. Further research may lead to practical applications for this approach.

Bibliography

Akhil, A. A., Huff, G., Currier, A. B., Kaun, B. C., Rastler, D. M., Chen, S. B., Cotter, A. L., Bradshaw, D. T., and Gauntlett, W. D. (2013). *DOE/EPRI 2013 Electricity Storage Handbook in Collaboration with NRECA.* SANDIA REPORT SAND2013-5131 (July 2013). 26

Allen, K. G., von Backström, T. W., Kröger, D. G., and Kisters, A. F. M. (2014). "Rock bed storage for solar thermal power plants: Rock characteristics, suitability, and availability." *Solar Energy Materials & Solar Cells* 126, pp. 170–183. DOI: 10.1016/j.solmat.2014.03.030. 73

Camargos, T. P. L., Pottie, D. L. F., Ferreira, R. A. M., Maia, T. A. C., and Porto, M. P. (2018). "Experimental study of a PH-CAES system: Proof of concept" *Energy* 165, pp. 630–638. DOI: 10.1016/j.energy.2018.09.109. 32, 33

Cava, F., Kelly, J., Peitzke, W., Brown, M., and Sullivan, S. (2016). "Advanced rail energy storage: green energy storage for green energy." In Letcher, T. M., ed. *Storing Energy - with Special Reference to Renewable Energy Sources.* Amsterdam: Elsevier, pp. 69–86. DOI: 10.1016/B978-0-12-803440-8.00004-X. 21

Choudhury, C., Chauhan, P. M., and Garg, H. P. (1995). "Economic design of a rock bed storage device for storing solar thermal energy." *Solar Energy* 55, pp. 29–37. DOI: 10.1016/0038-092X(95)00023-K. 56

Crawley, G. M., ed. (2017). *Energy Storage.* Singapore: World Scientific. DOI: 10.1142/10420.

Ding, Y. L., Tong, L. G., Zhang, P. K., Li, Y. L., Radcliffe, J., and Wang, L. (2016). "Liquid air energy storage." Chapter 9 in Letcher, T. M., ed., *Storing Energy - with Special Reference to Renewable Energy Sources.* Amsterdam: Elsevier, pp. 167-181. DOI: 10.1016/B978-0-12-803440-8.00009-9.

Dunlap, R. A. (2019). *Sustainable Energy*, 2nd ed. Boston, MA: Cengage.

Energiasalv Pakri OÜ (2010). "Brief description of the Muuga seawater-pumped hydro accumulation power plant." *Project ENE 1001.* Available at http://energiasalv.ee/wp-content/uploads/2012/07/Muuga_HAJ_17_02_2010_ENG.pdf. 15

Fujihara, T., Imano, H., and Oshima, K. (1998). "Development of pump turbine for seawater pumped storage power plant." *Hitachi Review* 47, pp. 199–202. Available at http://www.new4stroke.com/salt%20water%20pumped%20storage.pdf. 12

Gravity Power LL, www.gravitypower.net.

Guney, M. S. and Tepe, Y. (2017). "Classification and assessment of energy storage systems." *Renewable and Sustainable Energy Reviews* 75, pp. 1187–1197. DOI: 10.1016/j. rser.2016.11.102. 35

Gunnlaugsson, E. (2003). "Reykjavik energy -District heating in Reykjavik and electrical production using geothermal energy." *The United Nations University Geothermal Training Program - IGC2003 Short Course,* pp. 67–78, available at https://orkustofnun.is/gogn/ unu-gtp-report/UNU-GTP-2003-01-06.pdf. 58

Hahn, H., Hau, D., Dick, C., and Puchta, M. (2017). "Techno-economic assessment of a subsea energy storage technology for power balancing services." *Energy* 133, pp. 122–125. DOI: 10.1016/j.energy.2017.05.116. 16

Howari, F. M., Sadooni, F. N., and Goodell, P. C. (2008). "Assessment of water bodies of United Arab Emirates coastal Sabkhas as potential sites for natural salinity gradient solar ponds." *Journal of Energy Engineering* 134 pp. 111–120 DOI: 10.1061/(ASCE)0733-9402(2008)134:4(111). 68

Huggins, R. (2016). *Energy Storage: Fundamentals, Materials and Applications,* 2nd ed. Berlin: Springer.

Kabbara, Joseph, A., M., Groulx, D., Aldred, P., and White, M. A (2016). "Characterization and real-time testing of phase-change materials for solar thermal energy storage." *International Journal Energy Research* 40, pp. 61–70. DOI: 10.1002/er.3336. 76

Khalilian, M. (2017). "Energetic performance analysis of solar pond with and without shading effect." *Solar Energy,* 157 pp. 860–868. DOI: 10.1016/j.solener.2017.09.005. 65

Leblanc, J., Akbarzadeh, A., Andrews, J., Lu, H., and Golding, P. (2011). "Heat extraction methods from salinity-gradient solar ponds and introduction of a novel system of heat extraction for improved efficiency." *Solar Energy* 85, pp. 3103–3142. DOI: 10.1016/j. solener.2010.06.005. 66, 67

Letcher, T. M., ed. (2016). *Storing Energy - with Special Reference to Renewable Energy Sources,* Amsterdam: Elsevier.

Liu, H. and Jiang, J. (2007). "Flywheel energy storage—An upswing technology for energy sustainability." *Energy and Buildings* 39, pp. 599–604. DOI: 10.1016/j.enbuild.2006.10.001.

Lu, H., Walton, J. C., and Hein, H. (2002). "Thermal desalination using MEMS and salinity-gradient solar pond technology." University of Texas at El Paso Cooperative Agreement No. 98-FC-81-0047, Desalination Research and Development Program Report No. 80, August 2002, U.S. Department of the Interior, Bureau of Reclamation, Technical Service Center. 70

Luo, X., Wang, J., Dooner, M., and Clarke, J. (2015). "Overview of current development in electrical energy storage technologies and the application potential in power system operation." *Applied Energy* 137, pp. 511–536. DOI: 10.1016/j.apenergy.2014.09.081. 3

Malik, N., Date, A., Leblanc, J., Akbarzadeh, A., and Meehan, B. (2011). "Monitoring and maintaining the water clarity of salinity gradient solar ponds." *Solar Energy* 85, pp. 2987–2996. DOI: 10.1016/j.solener.2011.08.040. 69

Sebastián, R. and Peña-Alzola, R. (2012). "Flywheel energy storage systems: Review and simulation for an isolated wind power system." *Renewable and Sustainable Energy Reviews* 16, pp. 6803–6813. DOI: 10.1016/j.rser.2012.08.008. 40

Singh, H., Saini, R. P., and Saini, J. S. (2013). "Performance of a packed bed solar energy storage system having large sized elements with low void fraction." *Solar Energy* 87, pp. 22–34. DOI: 10.1016/j.solener.2012.10.004. 56

Singh, P. L., Deshpandey, S. D., and Jena, P. C. (2015). "Thermal performance of packed bed heat storage system for solar air heaters." *Energy for Sustainable Development* 29, 112–117. DOI: 10.1016/j.esd.2015.10.010. 57

Slocum, A. H., Fennell, G. E., Dundar, G., Hodder, B. G., Meredith, J. D. C., and Sager, M. A. (2013). "Ocean renewable energy storage (ORES) system: Analysis of an undersea energy storage concept." *Proc. IEEE* 101, pp. 906–924. DOI: 10.1109/JPROC.2013.2242411.

Wang, H., Wang, L., Wang, X., and Yao, E. (2013). "A novel pumped hydro combined with compressed air energy storage system." *Energies* 6 1554-1567. DOI:10.3390/en6031554. 30

Yao, E., Wang, H., Liu, L., and Xi, G. (2015). "A novel constant-pressure pumped hydro combined with compressed air energy storage system." *Energies* 8, pp. 154–171 DOI:10.3390/en8010154. 29, 31

Renewable Energy

Volume 3: Electrical, Magnetic, and Chemical Energy Storage Methods

Richard A. Dunlap
Dalhousie University

SYNTHESIS LECTURES ON RENEWABLE ENERGY TECHNOLOGIES #7

ABSTRACT

This volume considers various methods of energy storage that make use of electrochemical reactions, electric and magnetic fields, and chemical reactions. This book begins with a consideration of the use of batteries as a means of storing electrical energy. Various common battery chemistries are presented along with a summary of common battery sizes. The electrochemistry of a lithium-ion (Li-ion) cell is discussed in detail. Sodium-based batteries are discussed, as are vanadium flow batteries. The applications of batteries for energy storage are overviewed, concentrating on transportation technologies and grid-scale storage. Methods for storing energy in the form of electric fields include the use of supercapacitors and superconducting coils. The design of capacitors, including supercapacitors, pseudocapacitors, and hybrid capacitors is presented. The applications of supercapacitors for high-power, short-term energy storage are discussed. The use of superconducting magnets to store large amounts of electrical energy without resistive loss is presented. The application of superconducting electrical storage for grid stability is considered. Final chemical energy storage techniques are considered. The use of hydrogen as an energy carrier is discussed in detail. The concept of a future hydrogen economy has been popular in recent years. This volume considers the efficiency of such an approach. Other chemical energy carriers, such as methane, methanol, and ammonia, are discussed.

KEYWORDS

renewable energy, sustainability, energy storage technology, climate change, alternative energy

Contents

Preface

Due to diminishing fossil fuel resources and the adverse environmental impact of their continued use, it is essential that carbon-free renewable energy sources are developed. Most renewable energy resources, such as solar, wind, hydroelectric, and geothermal, are not constant in time and/or are not portable. For this reason, the development of renewable energy must be accompanied with the development of energy storage capabilities in order to provide energy when it is needed and for portable applications.

The present volume considers some of the important technologies for energy storage. Chapter 1 deals with batteries, which utilize electrochemical reactions to store electrical energy and are one of the most important energy storage technologies with applications that range from consumer electronics to transportation and grid-scale storage. Chapter 2 considers the storage of energy in electric and magnetic fields through the use of supercapacitors and superconducting magnets, respectively. Finally, Chapter 3 looks at some chemical approaches to energy storage. Electrical energy may be used to produce hydrogen and this energy can then be recovered by combustion or through the use of fuel cells. In addition, hydrogen may be used to produce methane, methanol, or ammonia, and these may be used as fuels in fuel cells or internal combustion engines. The various advantages and disadvantages of these energy carriers are discussed.

CHAPTER 1

Batteries

1.1 INTRODUCTION

Batteries are probably the best-known energy storage device. We encounter them in a wide variety of applications in our daily lives. They power electronic devices such as watches, cell phones and notebook computers and are used to start gasoline and diesel-powered vehicles. Batteries store energy using electrochemical reactions and a large fraction of those that we encounter utilize reactions that are reversible, meaning that the batteries can be recharged. The present chapter gives an overview of the common types of batteries that are in use, as well as some types of batteries that are in the early stages of development and commercialization. It also provides a brief description of the physics and chemistry of how batteries work. Finally, some important battery applications are presented.

1.2 TYPES OF BATTERIES

The terms "cell" and "battery" are sometimes used ambiguously or interchangeably, but both have fairly well-defined meanings. A cell is an electrochemical unit containing two electrodes and an electrolyte that provides a means of storing energy by electrochemical reactions. A battery is one or more electrochemical cells with appropriate packaging so as to constitute a practical device that is suitable for use in applications. In the present section, batteries are considered in terms of their overall characteristics. In the next section cells, with particular reference on Li-ion cells, are discussed in terms of the fundamental physics and chemistry of their operation.

Batteries may be categorized in different ways. One approach is to divide batteries into two categories: non-rechargeable and rechargeable. Non-rechargeable batteries are referred to as primary batteries and rechargeable batteries are referred to as secondary batteries. Another approach to categorizing batteries is on the basis of their chemistry. Different chemistries include, for example, carbon-zinc batteries, Li-ion batteries, and lead-acid batteries.

1.2.1 BATTERY CHEMISTRY

There are numerous different chemistries that can be used to produce batteries. Some of the most common are shown in Table 1.1. Carbon-zinc and alkaline are typically only used as primary batteries, while the other chemistries can be used to produce secondary batteries. The voltages given are for a single cell. More than one cell can be connected in series inside the battery to provide larger voltages. The specific energy refers to the energy produced per unit mass of the battery. This depends mainly on the chemistry, but also on the details of battery construction. The specific energy is also a function of battery size, as smaller batteries tend to have lower specific energy because the non-active components tend to account for a greater proportion of the mass. The values given in the table are typical of larger size batteries of each particular type.

Table 1.1: Typical properties of some common battery chemistries (NiMH = nickel metal hydride)

Chemistry	Cell Voltage (V)	Specific Energy (mWh/g)
Carbon-zinc	1.5	55
Alkaline	1.5	160
Pb-acid	2.1	36
Ni-Cd	1.2	56
NiMH	1.2	78
Li-ion	3.6	160

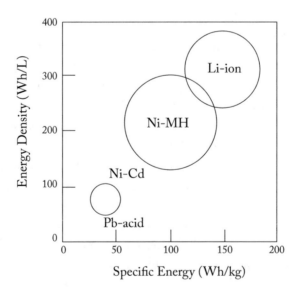

Figure 1.1: Ranges of energy density as a function of specific energy for some secondary battery chemistries.

From the rechargeable batteries shown in Table 1.1, it is clear that Li-ion batteries have the largest specific energy. This feature is also illustrated in Figure 1.1, which shows a graph of the energy density, that is the energy per unit volume, as a function of the specific energy. The ranges on the graph for the different battery chemistries are shown. It is clear that in many devices where size and weight are important factors, e.g., cell phones, iPads, etc., and in applications where a large energy storage capacity and portability is required, e.g., battery electric vehicles, Li-ion batteries are used almost exclusively. Other battery chemistries may be more cost effective than Li-ion, and in cases where small size and weight and/or large capacity is not as crucial a factor, other battery chemistries are also commonly used, for example, Pb-acid for starting gasoline engines, NiMH for hybrid vehicles, and Ni-Cd for power tools.

1.2.2   BATTERY SIZES

Batteries come in a wide variety of sizes. The most common sizes used for primary batteries are shown in Table 1.2. The capacity, shown in mAh in the table, is discussed further below. The same size designations are often used for rechargeable Ni-Cd and NiMH batteries.

Table 1.2: Common sizes of carbon-zinc and alkaline primary batteries showing their dimensions, mass, and capacity (in mAh)

Name	Diameter (mm)	Length (mm)	Carbon-Zinc		Alkaline	
			Mass (g)	Capacity (mAh)	Mass (g)	Capacity (mAh)
AAAA	8.3	42.5	-	-	6	625
AAA	10.5	44.5	10	540	12	1,200
AA	14.5	50.5	19	1,100	23	2,700
C	26.2	50	48	3,800	66	8,000
D	34.2	61.5	98	8,000	135	12,000

Other size batteries do not necessarily have a standard numbering scheme. In fact, batteries with the same specifications from different manufactures may have different numbers. Some lithium batteries, however, use a reasonably standard method of numbering. Cylindrical Li-ion batteries are given a four- or five-digit number. The number refers to the size of the battery. The first two digits of the number refer to the diameter of the battery in mm. The last two or three digits of the number refer to the length (or thickness) of the battery in tenths of a mm. Two examples are shown in Figure 1.2. The 2032 battery is 20 mm in diameter and 3.2 mm thick. The 18650 battery is 18 mm in diameter and 65.0 mm long.

Figure 1.2: Two examples of common cylindrical Li-ion batteries: 2032 (left) and 18650 (right) (image courtesy of Richard A. Dunlap).

2.2.3   BATTERY COST

Batteries are a convenient portable energy storage medium. While rechargeable (secondary) batteries can be used repeatedly by recharging, typically from grid-powered adapters, non-rechargeable (primary) batteries can only be used once. While primary batteries are generally less expensive than secondary batteries, they are not economical for use in devices that require a fairly continuous supply of energy over a long period of time. These applications include cell phones, notebook computers, and battery electric vehicles. To understand the cost of electricity from primary batteries we can look at Table 1.2. AA batteries are in common use for devices such as flashlights. From the table we see that a carbon-zinc AA battery has a capacity of 1,100 mAh. To calculate the energy content of the battery we use Ohm's law for power

$$P = VI.$$ (1.1)

Multiplying through by time gives an equation for energy

$$E = Pt = IVt,\qquad(1.2)$$

or in the units used here:

$$mWh = mAh \times volts.\qquad(1.3)$$

Since the battery voltage is about 1.5 V then the energy content will be

$$(1100 \text{ mAh}) \times (1.5 \text{ V}) = 1.65 \text{ Wh},\qquad(1.4)$$

or 0.00165 kWh. A package of four carbon-zinc AA batteries might cost about \$1 at a discount store, that is, about \$0.25 each. Thus, the cost of energy from the AA battery is

$$(\$0.25)/(0.00165 \text{ kWh}) = \$150 \text{ per kWh}.\qquad(1.5)$$

This can be compared to the price of electricity from typical public utility companies of about \$0.15 per kWh. So, the electricity from the battery costs about 1,000 times as much per unit energy as the electricity from the grid; see Figure 1.3. This stresses the point that, for applications such as battery electric vehicles, rechargeable batteries are not only convenient, they are the only economical choice.

Figure 1.3: Flashlight powered by a carbon-zinc AA battery at a cost of \$150 per kWh (image courtesy of Richard A. Dunlap).

1.3   HOW BATTERIES WORK

Batteries operate by transporting charge from one electrode to the other. In a secondary battery the reaction is reversible and charge transport is in one direction during charge and in the opposite direction during discharge. The details of this process depend on the details of the battery chemistry. Since Li-ion batteries are the most useful for many of the applications discussed in Section 1.5, we consider the details of the construction and operation of a Li-ion cell in the present section.

A Li-ion cell consists of two electrodes separated by an insulating material soaked with a lithium salt electrolyte dissolved in an organic solvent. The cell can produce a current of electrons that can do work on an external circuit by transporting Li-ions from one electrode at a lower potential to an electrode at a higher potential. The positive electrode is referred to as the cathode and the negative electrode is referred to as the anode. This definition refers to the direction of flow of the positive Li-ions and is opposite to the definition of cathode and anode in many other systems (such as photomultiplier tubes) where the charge flow is viewed from the perspective of negatively charge electrons. Typical electrode materials are a lithium metal oxide for the positive electrode and carbon (graphite) for the negative electrode. Some common electrode materials are shown in Table 1.3 (for cathodes) and Table 1.4 (for anodes). The potential of each material relative to lithium metal is shown in the tables.

Table 1.3: Potential of some cathode (positive electrode) materials relative to lithium metal

Material	Potential (V)
$LiCoO_2$	3.90
$LiFePO_4$	3.45
$LiMn_2O_4$	4.05

Table 1.4: Potential of some elemental anode (negative electrode) materials relative to lithium metal

Element	Capacity (mAh/g)	Lithium Compound	Li:element	Potential (V)
Al	992	LiAl	1:1	0.3
C	372	LiC6	0.167:1	0.1
Sb	660	Li_3Sb	3:1	0.9
Si	3,578	$Li_{15}Si_4$	3.75:1	0.4
Sn	992	$Li_{22}Sn_5$	4.4:1	0.5

Note that, in all cases, both electrode materials have a potential above elemental lithium where the positive electrode material is approximately 4 V and the negative electrode material is approximately 0.5 V. An energy level diagram is illustrated in Figure 1.4. The direction of flow of the Li$^+$ ions for the charge and discharge processes are shown.

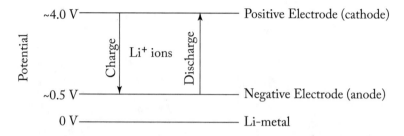

Figure 1.4: Potential of the positive and negative electrodes of a Li-ion battery relative to lithium metal. The direction of the flow of Li$^+$ ions during charge and discharge is shown.

The basic design of a Li-ion cell is shown in Figure 1.5. The electrodes are electrically isolated from one another by an insulating microporous separator typically made of polyethylene or polypropylene. The electrodes are electrically connected to conductive foils (the current collectors) which are typically aluminum for the positive cathode and copper for the negative anode. The electrodes are connected to the current collectors using a binder such as polyvinylidene fluoride with a conductive additive, usually carbon. The electrolyte consists of a lithium salt such as LiPF, LiBF, or LiClO dissolved in an organic solvent such as ethylene carbonate, dimethyl carbonate or diethyl carbonate. The lithium in the electrolyte acts as a pathway for the transport of Li-ions.

Li-ion batteries charge and discharge by exchanging lithium between the positive and negative electrodes. This type of battery is often referred to as a *rocking chair battery*, as the Li-ions rock back and forth between the electrodes. Figure 1.5 shows the direction of flow of the Li$^+$ ions during charge and discharge. The Li$^+$ ions flow through the electrolyte from one electrode to the other. Negative electrons, which are liberated from the Li$^+$ ions, travel through an external circuit. During charge, a current source does work on the charges and moves the electrons through the circuit. During discharge, the electrons do work on the external circuit.

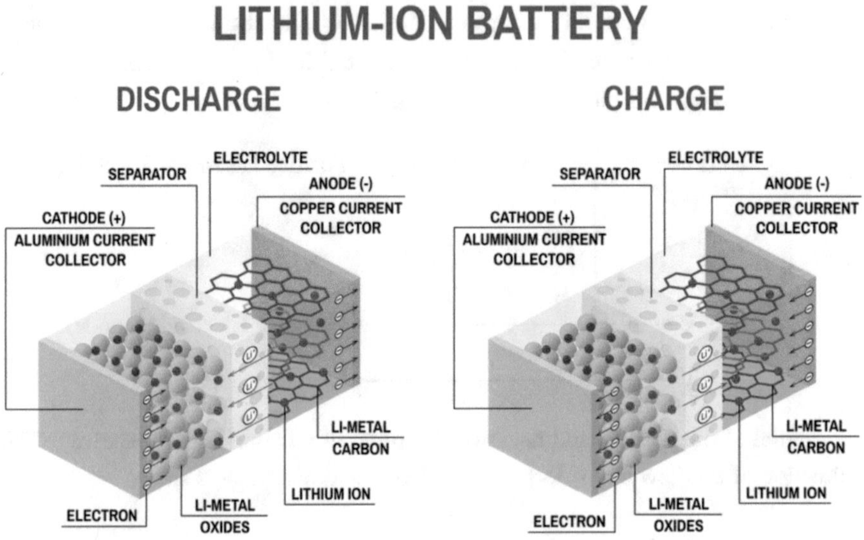

LITHIUM-ION BATTERY

DISCHARGE

CHARGE

Figure 1.5: (left) Flow of Li⁺ ions during the discharge process and (right) flow of the Li⁺ ions during the discharge process. The negative electrode (anode) is typically made from carbon and the positive electrode (cathode) is typically made from $LiCoO_2$. Image from Shutterstock.com, https://www.shutterstock.com/image-illustration/liion-battery-diagram-rechargeable-which-lithium-791697073?src=bDB-KCqfvuWF5VzmkxcIjA-1-15.

Typically, the lithium metal oxide positive electrode has a layered or tunneled structure so that the lithium can easily be accommodated in the crystal structure without significant structural changes to the host material. Similarly, typical carbon negative electrodes are carbon in a layered graphite-like structure where the lithium atoms can enter into the spaces between the carbon layers without significantly affecting the layered carbon structure. Such insertion of atoms (e.g., lithium) between the layers of a layered host material (e.g., graphite) is referred to as intercalation.

Using $LiMO_2$ and carbon as typical examples of positive and negative electrode materials, the reactions at the positive and negative electrodes during cell cycling are given in Figure 1.6. These reactions are referred to as oxidation-reduction reactions, often called redox reactions, and are reversible, proceeding in one direction during charge and the opposite direction during discharge, as shown in the figure.

Positive Electrode: $LiCoO_2$ $\xrightarrow[\text{Discharge}]{\text{Charge}}$ $Li_{1-x}CoO_2 + xLi^+ + xe^-$

Negative Electrode: $C + yLi^+ + ye^-$ $\xrightarrow[\text{Discharge}]{\text{Charge}}$ Li_yC

Overall: $LiCoO_2 + \frac{x}{y}C$ $\xrightarrow[\text{Discharge}]{\text{Charge}}$ $\frac{x}{y}Li_yC + Li_{1-x}CoO_2$

Figure 1.6: Charge and discharge reactions at the positive and negative electrodes in a typical Li-ion cell, along with the overall reaction for the cell.

The details of the reactions that take place at the anode and cathode can be most easily studied using Li-ion half cells. These are cells constructed with one electrode of either the anode or cathode material and the other electrode of pure lithium. Thus, either a positive electrode or a negative electrode material is referenced to metallic lithium. Either of these electrode materials is at a more positive potential than the lithium metal electrode. The charge and discharge processes for half-cells containing positive and negative electrodes are shown in Figure 1.7.

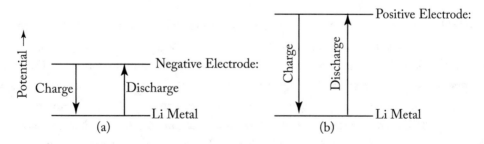

Figure 1.7: Potentials of charge and discharge reactions for half cells consisting of: (a) lithium metal vs. a negative electrode material (e.g., carbon); and (b) lithium metal vs. a positive electrode material (e.g., $LiCoO_2$).

As an example, we consider a Li–ion half cell made from a carbon electrode and a lithium metal electrode, as shown in Figure 1.7(a). As clearly shown in the figure, a cell constructed from a carbon (negative) electrode and a lithium metal electrode will be in the charged state. The initial discharge involves transporting lithium from the lithium metal electrode to the carbon electrode.

The voltage of the cell is determined by the quantity of lithium that has intercalated into the carbon electrode and is shown in Figure 1.8 during the initial discharge.

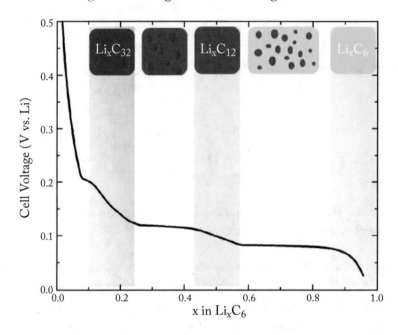

Figure 1.8: Voltage of a carbon vs lithium cell during the initial discharge. Based on Gallagher et al., 2012. Reprinted with permission from the Electrochemical Society.

It is clear that the voltage as a function of the lithium content of the graphite electrode does not follow a linear dependence but has steps, referred to as stages. These stages correspond to the formation of different lithium microstructures within the layered graphite structure. The structure of pure graphite consists of layers of hexagonal planes of atoms, as shown in Figure 1.9. The staging corresponds to the formation of planes of lithium atoms separated by planes of graphite, as shown in Figure 1.10. It is seen that stage n corresponds to a structure with lithium layers separated by n graphite layers. These types of studies provide considerable insight into the reactions that take place during the operation of an electrochemical cell.

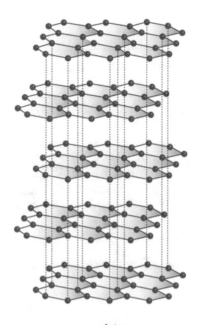

graphite
hexagonal crystal
layer distance : 335 pm

Figure 1.9: The layered hexagonal structure of graphite. Image from Shutterstock.com, https://www.shutterstock.com/image-illustration/diamond-graphite-fullerene-339037571?src=KCcScu8B-PoK1QcPLi0g8cg-1-9.

The value of x on the horizontal axis of Figure 1.8 is a measure of the number of Li^+ ions that have been transferred from the Li-counter electrode to the carbon electrode. This represents a quantity of charge and is referred to as the capacity. It is most common to express this as a charge per unit mass of electrode material (often called the specific capacity or gravimetric capacity). As a current is a charge per unit time, then a charge is a current multiplied by time. The common units of charge are mAh and the specific capacity will, therefore, be in mAh/g, as in Table 1.4 for different anode materials. The total capacity of a battery is, therefore, given in mAh, as shown for some primary batteries in Table 1.2.

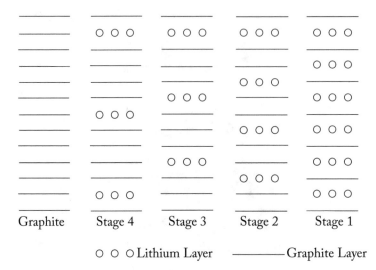

Figure 1.10: Graphical depiction of the different stages of lithium intercalation into graphite.

1.4   OTHER BATTERY TECHNOLOGIES

In addition to the battery chemistries discussed above, which are currently used commercially, there are numerous other approaches to battery chemistry and design that have met with some success or are in the development stages. However, batteries are not perfect for all applications. While different battery types are the most suitable for different applications, there are concerns that still need to be addressed for many of these applications. In many cases, batteries are too heavy or occupy too much volume for the capacity that they provide. In other cases, faster charge times would be beneficial. From an economic standpoint, reducing cost without compromising performance is always desirable. Two approaches to battery design that are particularly notable in their attempt to deal with these deficiencies are batteries based on sodium chemistry and vanadium flow batteries. These are discussed in the present section.

1.4.1   SODIUM BATTERIES

Li-ion batteries have been the most successful for many demanding applications which require high specific capacity, reasonably fast recharge times, and good cyclability. These include applications such as personal electronics, power tools, and, probably most importantly, battery electric vehicles and grid energy storage. Two concerns about the widespread use of lithium batteries are: (1) the high cost of lithium and (2) the limited world supply of lithium. In addition, the availability and cost of cobalt which is used in the cathode of many Li-ion batteries is also a concern. There is, therefore, the desire to produce batteries which perform as well as lithium batteries, but which utilize materials that are less expensive and more plentiful than lithium. If one looks at the periodic table, one sees that sodium falls directly below lithium, meaning that its outer electronic structure is the same as lithium's and its chemical properties are similar. Sodium is very plentiful (as part of all the NaCl on earth) and is, therefore, quite inexpensive. In recent years, there has been considerable research on the development of batteries that use sodium chemistry rather than lithium chemistry. Two possible approaches to the fabrication of batteries which use reactions involving sodium are discussed in the next sub-sections.

Sodium-Sulfur Batteries

Sodium-sulfur batteries are a type of molten salt battery which use electrodes consisting of liquid sodium and liquid sulfur. The general design of the sodium-sulfur cell is illustrated in Figure 1.11. Electrical connections are made with liquid sodium, which acts as the anode, and with the liquid sulfur, which acts as the cathode, by an inner metal can and the outer metal cell container, respectively. The liquid sodium and liquid sulfur are separated by a solid membrane consisting of beta-alumina (called the beta-alumina solid electrolyte or BASE). The BASE is selectively permeable to Na^+ ions which, during discharge, travel from the sodium electrode to the sulfur electrode, while electrons travel through an external circuit to be recombined with the Na-ions at the cathode. The BASE membrane is relatively impermeable to electrons and this largely eliminates self-discharge of the charged cell. The overall reaction during discharge is given as

$$2Na + 4S \rightarrow Na_2S_4. \tag{1.6}$$

This reaction is reversible and, consequently, the cell is rechargeable. Some advantages of sodium-sulfur batteries over many other types of battery are:

- they have high energy density,

- they have long cycle life,

- they have high charge/discharge efficiency, and

- they are fabricated from readily available inexpensive materials.

Unfortunately, sodium-sulfur batteries also have some drawbacks. First, the battery must be kept at a temperature of 300–350°C, in order to maintain the sodium and sulfur in the liquid state. This means that there is considerable additional equipment necessary to maintain the operating conditions for the battery. Second, the sodium polysulfide that is formed in the sulfur electrode as a result of the diffusion of Na-ions during discharge is highly corrosive. This means that the cell container must be fabricated from a metal, e.g., chromium or molybdenum, that resists corrosion from the sodium polysulfide.

Because of the specific operating conditions that are necessary for sodium-sulfur batteries, they are most applicable to large-scale, stationary facilities. As discussed below, these include stand-alone, off-grid systems, and grid energy storage. They provide an attractive alternative to, e.g., pumped hydroelectric storage, in cases where an appropriate geographical location is not available.

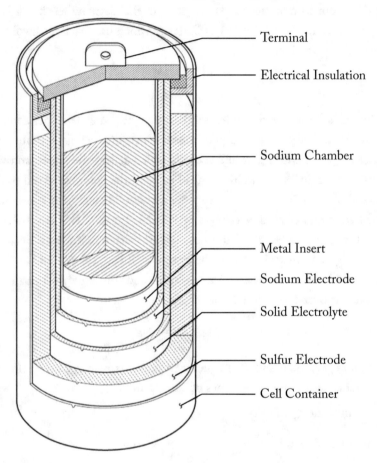

— Terminal

— Electrical Insulation

— Sodium Chamber

— Metal Insert

— Sodium Electrode

— Solid Electrolyte

— Sulfur Electrode

— Cell Container

Figure 1.11: Diagram of the internal structure of a sodium-sulfur battery; https://commons.wikimedia.org/wiki/File:NaS_battery.png.

Sodium-Ion Batteries

Na-ion batteries follow along the lines of Li-ion batteries as shown in Figure 1.5, except that sodium ions rather than lithium ions carry the positive charge between the anode and cathode during discharge. The general design of a Na-ion battery is illustrated in Figure 1.12. It is seen in the figure that the cathode is typically made from a layered oxide material, while the anode is made from a metal alloy or oxide or a carbon-based material. It is important to note that a major difference between Li-ion cells and Na-ion cells is related to the relative sizes of the Li-ion and the Na-ion. The

Na$^+$ ion is much larger than the Li$^+$ ion, having 10 electrons rather than 2, with an ionic radius of 0.102 nm compared to 0.076 nm. It is, therefore, desirable to utilize electrode materials that have considerable open space in their structures.

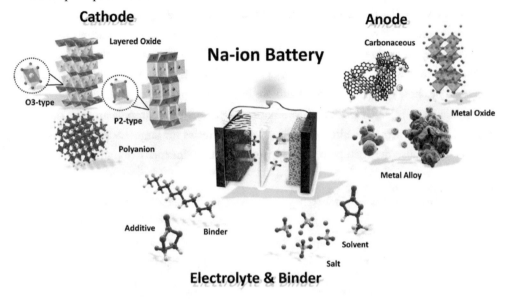

Figure 1.12: Design of a Na-ion battery (Hwang et al., 2017).

While much research has been conducted on Na-ion batteries, and much of this has been devoted to the development of anode and cathode materials, there is more work needed to produce commercially viable cells. To date, hard carbon is generally considered to be the most promising anode material. Hard carbon is a non-graphitic material which consists of a random arrangement of planes of carbon atoms, as shown in Figure 1.13, and which provides space for Na$^+$ ions to enter into the structure.

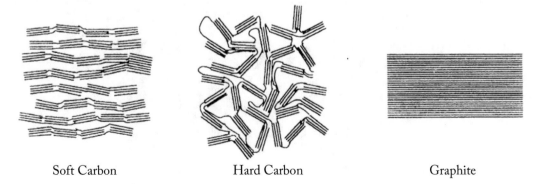

Figure 1.13: Structure of different forms of carbon. Based on Wakihara, 2001, reprinted with permission from Elsevier.

Layered metal oxides have shown the most promise as cathode materials for Na-ion batteries. Some typical materials of this type are illustrated in Figure 1.14. The figure shows the location of the sodium atoms which enter into the structure between metal oxide layers. The structures shown in the figure are designated O2, O3, P2 and P3. The letter indicates the local symmetry of the oxygen atoms around the sodium: "O" for octahedral and "P" for prismatic. The number indicates the ordering of the metal oxide layers, as shown Figure 1.14. The O3 and P2 structures have shown the most promise as cathode materials for Na-ion batteries. O3 materials such as $NaNi_{0.5}Mn_{0.5}O_2$ are particularly attractive, as they show good performance and are free from expensive cobalt.

While there is much work still to be done to perfect the materials and the design of Na-ion batteries, they are attractive alternatives to Li-ion batteries for large-scale systems, such as grid energy storage, because they can be produced from common, inexpensive materials.

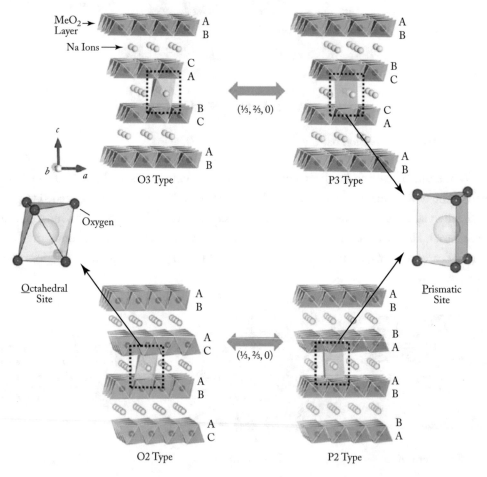

Figure 1.14: Crystal structures of layered compounds that are of interest as cathode materials for Na-ion batteries. Based on Yabuuchi and Komaba, 2014, reprinted with permission.

1.4.2   VANADIUM FLOW BATTERIES

Vanadium flow batteries, also known as vanadium redox batteries or vanadium redox flow batteries, are a type of battery that uses a liquid anode and cathode. The general design of a flow battery is shown in Figure 1.15. This battery utilizes the fact that vanadium can exist in solution in four different oxidation states; V^{2+}, V^{3+}, V^{4+}, and V^{5+}. The vanadium is in solution (typically in sulfuric acid) where ions in the V^{4+} and V^{5+} states exist in the cathode (positive electrode) and ions in the V^{2+} and V^{3+} states exist in the anode (negative electrode). The two electrode solutions are separated by a proton exchange membrane as illustrated in the figure. The reactions at the positive and negative electrodes during charge and discharge are shown in Figure 1.16.

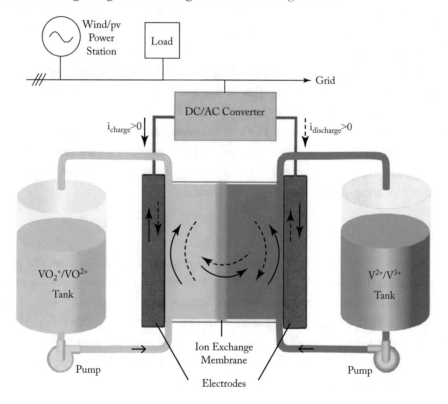

Figure 1.15: Diagram of a vanadium flow battery. Image based on Alotto et al., 2014, reprinted with permission from Elsevier.

$$\text{Positive Electrode} \quad VO^{2+} + H^2O \underset{\text{Discharge}}{\overset{\text{Charge}}{\rightleftharpoons}} VO_2{}^+ + 2H^+ + e^-$$

$$\text{Negative Electrode} \quad V^{3+} + e^- \underset{\text{Discharge}}{\overset{\text{Charge}}{\rightleftharpoons}} V^{2+}$$

Figure 1.16: Reactions at the positive and negative electrodes of a vanadium flow battery during charge and discharge.

During the charge process, V^{4+} (tetravalent vanadium), which exists as VO^{2+} ions in the positive electrode, is oxidized to V^{5+} (pentavalent vanadium). Hydrogen ions released from solution travel through the proton exchange membrane where they recombine with electrons that flow through the external circuit. At the negative electrode V^{3+} (trivalent vanadium) ions are reduced to V^{2+} (bivalent vanadium) ions. The reactions proceed in the opposite direction during discharge.

As illustrated in Figure 1.15, tanks of liquid electrode can be attached to the cell and pumps can continuously supply fresh solution to the cell. In this way batteries with very large capacities can be fabricated. It is also seen that the battery can be recharged by merely replacing the discharged solutions with fresh solution. This rapid approach to recharging the battery is attractive for many applications, particularly for vehicles, where recharge times are a concern. However, flow batteries are typically large and have a low specific energy capacity. As a result, they are most appropriate for stationary applications, such as grid storage.

1.5   BATTERY APPLICATIONS

Batteries are a part of our daily lives. From personal electronic devices such as cell phones and notebook computers that use rechargeable lithium batteries, to fossil fuel-powered vehicles that use Pb-acid batteries for starting or hybrid vehicles that use NiMH batteries for energy storage, we have become dependent on batteries as a means of storing electrical energy. In this section, we look at some ways in which batteries have been applied to the energy use of society. We concentrate on three aspects of battery use, electric vehicles, off-grid systems, and grid storage, as these have the most relevance to our implementation of renewable energy sources.

1.5.1   VEHICLES

Vehicles, that is road vehicles such as automobiles, present one of the most difficult challenges for energy storage. The energy storage method must have a sufficiently high specific energy and specific power so that a reasonable mass of storage material can provide a reasonable range and reasonable power for the vehicle. Recharge time and cost are also important factors.

Battery electric vehicles have a long history that dates back to the 19th century. From around 1880 until about 1920 they were quite popular, see Figure 1.17, and had a number of advantages over gasoline-powered vehicles. They were quieter, easier to start, more dependable, and more powerful than their fossil fueled counterparts. Around 1900 in North America, 22% of vehicles had internal combustion engines, 40% had steam engines, and 38% were electric.

Figure 1.17: Woods Electric Queen Victoria Brougham motor vehicle manufactured around 1905; https://commons.wikimedia.org/wiki/File:1906_Woods_Queen_Victoria_Electric.jpg.

Around 1920, the interest in electric vehicles gradually subsided as a result of the growing need to drive longer distances, along with improvements in gasoline engines, particularly the development of the electric starter. Interest in electric vehicles was renewed in the 1990s as a result of the growing awareness of the environmental effects of fossil fuels use, as well as the limitations of their long-term availability. One of the first electric vehicles offered to the general public during that era was the General Motors EV1, as shown in Figure 1.18. The first version of the EV1 (Generation I, model year 1997) used Pb-acid batteries and had a range of 130–160 km. Generation II EV1s (model year 1999) used NiMH batteries and had a range of 160–230 km. A significant number of automobile makers now produce battery electric vehicles (BEVs). Before we look at the current availability of these vehicles, it is of interest to analyze the energy and power requirements for BEVs.

Figure 1.18: General Motors EV1 electric vehicle made from 1996–1999; https://commons.wikimedia.org/wiki/File:EV1A014_(1).jpg. RightBrainPhotography (Rick Rowen), https://creativecommons.org/licenses/by-sa/2.0/.

Energy Requirements for Vehicles

We can understand the energy and power requirements of road vehicles by looking in detail at some recent model gasoline engine vehicles. Table 1.5 shows the specifications of some currently available passenger vehicles. Engine power in SI units is measured in kW. This is related to the convention used in many countries, where the power is measured in horsepower, by the relation

$$kW = 0.7457 \times hp. \tag{1.7}$$

The fuel consumption in L/100 km (which is used in many countries) is related to the value in miles per gallon (mpg) used in the U.S. as

$$\frac{L}{100 \text{ km}} = \frac{235}{mpg}. \tag{1.8}$$

The energy utilized pcr km of driving is determined from the fuel consumption and the energy content of gasoline (34.8 MJ/L). Thus, we may write

$$\frac{MJ}{km} = 0.348 \frac{MJ}{L} \times \frac{L}{100\ km}. \tag{1.9}$$

Table 1.5: Specifications of some current model (2018 model year) gasoline engine vehicles. The fuel consumption is the combined value for highway/city driving as determined by the U.S. Environmental Protection Agency

Make	Mode	Type	Mass (kg)	Engine Power (kW)	Fuel Consumption (L/100 km)	Range (km)	Energy Used (MJ/km)	Energy to wheels (MJ/km)
Nissan	Versa Note	Subcompact	1,116	81	6.9	591	2.4	0.48
Toyota	Corolla	Compact	1,288	98	7.3	680	2.5	0.50
BMW	530	Full size	1,699	185	8.7	782	3.0	0.65

The value from Equation (1.9), as given in the table, is the energy content of the average amount of gasoline burned per kilometer driven. It is, however, not the energy which is delivered to the wheels of the vehicle. The internal combustion gasoline engine is a heat engine that has a typical thermodynamic efficiency of around 20% (meaning that about 80% of the energy released during the combustion of the gasoline is lost to the environment as heat). Thus, the table gives the energy delivered to the wheels per kilometer driven as 20% of the energy content of the fuel. Table 1.5 shows the energy delivered to the wheels ranges from around 0.5 MJ/km for of a typical small passenger vehicle to 0.65 MJ/km for a family-size luxury vehicle. For the subsequent discussion in the present chapter, we use 0.55 MJ/km as a typical value for a light duty vehicle. As the energy capacity of batteries is typically expressed in units of kWh (see Equation (1.3)), then the energy requirement for vehicle propulsion may be expressed as

$$(0.55 MJ/km)/(3.6\ MJ/kWh) = 0.15\ kWh/km. \tag{1.10}$$

Since the battery/motor drive system is typically about 90% efficient the energy storage requirement is about

$$(0.15\ kWh/km)/(0.9) = 0.17\ kWh/km. \tag{1.11}$$

From Table 1.5 we see that various vehicle designs utilize energy at the wheels (converted to kWh/km) from about 0.15–0.20 kWh/km.

Current Battery Electric Vehicle Production

In recent years, a large fraction of world automobile manufacturers has produced battery electric vehicles. These include passenger vehicles, sports cars, and SUVs. Specifications of some of the more notable vehicles that have been available worldwide in recent years are given in Table 1.6. Photographs of these vehicles are shown in Figures 1.19–1.23. The values of the battery energy used per kilometer based on the battery capacity and the range is between 0.15 kWh/km and 0.20 kWh/km, in good agreement with the analysis based on gasoline-powered vehicles as previously mentioned.

Table 1.6: Specifications for some recent model battery electric vehicles. Many vehicles have different battery packs available; values are shown for the highest capacity battery pack commonly available to consumers. All vehicles use rechargeable Li-ion batteries. The range in the table is the U.S. EPA value (if available) or the manufacturer's value. The energy per kilometer is battery energy per km as given in Equation (1.11)

Make	Model	Power (kW)	Battery Energy Capacity (kWh)	Range (km)	Battery Energy per km (kWh/km)	Figure
BDY	e3	75	61	300	0.20	1.19
BMW	i3	125	33	183	0.18	1.20
Chevrolet	Bolt	150	60	383	0.16	1.21
Hyundai	Kona EV	150	64	418	0.15	1.22
Nissan	Leaf	110	40	243	0.16	1.23
Tesla	Model 3	211	75	500	0.15	1.24

Figure 1.19: **BDY e6 battery electric vehicle manufactured in China;** https://commons.wikimedia.org/wiki/File:Shenzhen_BYD_e-taxi_new_look_1.jpg.

Figure 1.20: **BMW i3 battery electric vehicle;** https://commons.wikimedia.org/wiki/File:2018_BMW_i3_facelift_(1).jpg.

Figure 1.21: Chevrolet Bolt battery electric vehicle. Image from Ed Aldridge, Shutterstock.com, https://www.shutterstock.com/image-photo/detroit-us-january-92017-chevrolet-bolt-559022086.

Figure 1.22: Hyundai Kona EV battery electric vehicle. Image from Grzegorz Czapski, Shutterstock.com, https://www.shutterstock.com/image-photo/poznan-poland-april-05-2018-metallic-1066724234?src=3gWYG_6UENqkUMnvNRpgpw-1-2.

Figure 1.23: **Nissan Leaf battery electric vehicle.** Image from Anton Ukolov, Shutterstock.com, https://www.shutterstock.com/image-photo/25-june-2017-sydney-australia-electric-666547597?src=F-c8FleqCPQAay87SFMZqrQ-1-6.

Figure 1.24: **Tesla Model 3 battery electric vehicle;** https://commons.wikimedia.org/wiki/File:Tesla_Model_3_Monaco_IMG_1212.jpg.

The analysis above provides a fairly accurate measure of a battery electric vehicle's range on the basis of battery capacity. It should be noted, however, that the energy consumption per km, and, hence, the range is highly dependent on conditions. Particularly, temperature is an important factor in determining the range of a battery electric vehicle. This feature comes about because of two factors. First, at temperatures above or below ambient temperature, it is likely that air conditioning or heating, respectively, will be required, and these must be provided by energy from the battery. The second important factor relates to the temperature dependence of the energy storage capacity of a battery. An example of this effect is illustrated in Figure 1.25. As an example of the effects of outside temperature and the use of air conditioning or heating on vehicle range, the data provided by Tesla for the Model S with the 90D battery pack (90 kWh capacity) is illustrated in Figure 1.26. It is seen that the vehicle is most efficient at 20°–30°C. There is a slight decrease in efficiency at higher temperatures which results from the need for air conditioning, but this is partially off-set by increased battery capacity. There is a significant decrease in efficiency at low temperatures which results for a combination of the need for heating and the decreasing battery capacity.

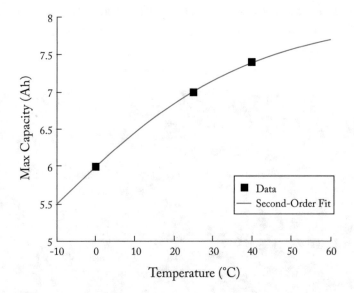

Figure 1.25: Temperature dependence of the capacity of a 6 Ah Li-ion battery as a function of temperature. Based on Johnson et al., 2001.

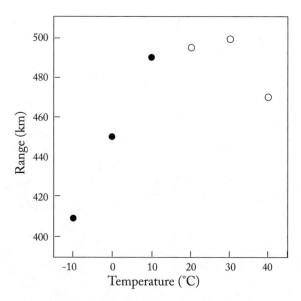

Figure 1.26: Estimated range of the Tesla Model S with a 90 kWh battery as a function of outside temperature. Solid data points are with heater on and open data points are with air conditioner on. Speed is a constant 96 km/h (60 mph).

Battery electric vehicle sales have grown substantially since the first modern era BEVs were produced in the 1990s. Figure 1.27 shows the estimated total number of electric vehicles in the U.S. as a function of year. The most recent growth period started around 2010. In recent years, there has also been growth in the number of battery electric vehicles worldwide as illustrated in Figure 1.28. This trend is continuing, and it is estimated that as of 2018 there are approximately 750,000 BEVs in the U.S. and 3.2 million worldwide. These numbers can be compared to the total number of road vehicles in the U.S., 250 million, and worldwide, 1.2 billion. In both cases, BEVs represent about 0.3% of the total. Models for the growth of battery electric vehicles predict that they will represent about half of all vehicles somewhere around 2045–2050.

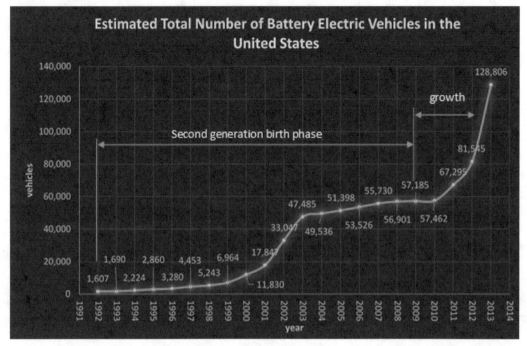

Figure 1.27: Total number of battery electric vehicles registered in the U.S. as estimated by the U.S. Energy Information Administration; https://commons.wikimedia.org/wiki/File:Estimated_Total_Number_of_Battery_Electric_Vehicles_in_the_United_States.jpg, https://creativecommons.org/licenses/by-sa/3.0/deed.en.

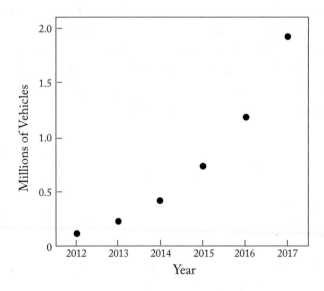

Figure 1.28: Estimated number of battery electric vehicles worldwide as a function of year.

1.5.2   OFF-GRID AND EMERGENCY POWER SYSTEMS

Off-grid power systems that utilize battery storage include stand-alone residential systems that obtain power from solar photovoltaics or wind turbines. It is important to note that the output from photovoltaic panels is direct current (DC), while the output from a wind turbine is typically alternating current (AC). Therefore, the integration of these power sources into domestic power with battery storage must take these differences into account. Figure 1.29 shows a simple diagram of a system which utilizes a battery as storage for a stand-alone photovoltaic system. A system for battery storage of power from a wind turbine is illustrated in Figure 1.30. The diagrams show how the source determines the way in which rectifiers and inverters are incorporated into the circuits.

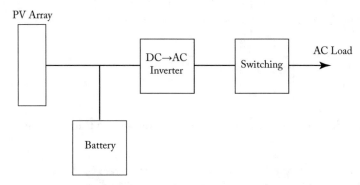

Figure 1.29: Simple schematic for the use of solar photovoltaics with battery storage for residential power.

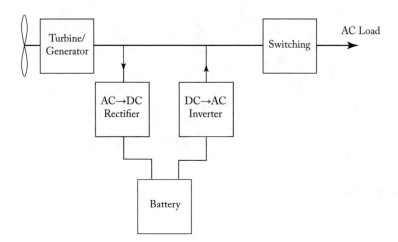

Figure 1.30: Simple schematic for the use of a wind turbine with battery storage for residential power.

Batteries offer a convenient means of providing emergency power in case of a grid power outage. A simple system for this purpose is shown in Figure 1.31. Such systems are in common use as uninterruptible power supplies (UPS) for personal computers. On a larger scale they are in common use for facilities that require an uninterrupted supply of power in the event of a grid outage. Such facilities include, most notably, data storage facilities and communication centers, as shown in Figure 1.32. In many cases, battery systems maintain power until backup generators can be brought on line.

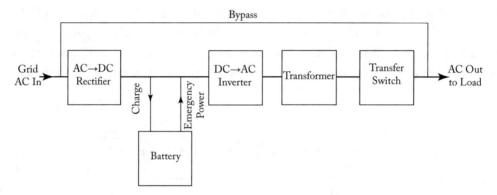

Figure 1.31: Schematic of a simple emergency power backup system using batteries.

Figure 1.32: Battery backup for cellular communication facility. Image from Shutterstock.com, https://www.shutterstock.com/image-photo/storage-battery-switchboard-cellular-communication-14510314?src=vAzXjHuNURINVVhSLSLdRA-1-89.

1.5.3   GRID ENERGY STORAGE

Grid energy storage is important as a means of grid stabilization to improve the quality of grid power, as a means to cover periods of peak demand and as a means of effectively utilizing renewable energy sources. At present, the largest grid battery storage facilities utilize sodium-sulfur batteries as given in Table 1.7. The world's largest battery storage facility is the Bruzen Substation in Japan.

Table 1.7: World's largest sodium-sulfur battery grid storage facilities

Name	Country	Energy Storage Capacity (MWh)	Power (MW)	Duration (h)
Buzen Substation	Japan	300	50	6
Rokkasho Aomori	Japan	245	34	7

Most grid-scale battery storage facilities at present use Li-ion batteries as summarized in Table 1.8. Figure 1.33 shows an illustration of a typical facility. The world's largest lithium-ion storage facility is the Hornsdale Power Reserve near Jamestown in Southern Australia. This facility is associated with the Hornsdale Wind Farm and is used to store electrical energy produced by the wind farm for distribution to the grid. It was constructed by Tesla, Inc. at a cost of about $50 million USD, and uses cylindrical 21700 Li-ion cells (21 mm diameter by 70.0 mm long). These are the same cells that Tesla produces for use in the Tesla Model 3 battery electric vehicle described above.

Table 1.8: World's largest Li-ion battery grid storage facilities. Note: the Hornsdale battery facility is in two sections, one which provides 70 MW output for a short period of time and the other which provides 30 MW for an extended period of time

Name	Country	Energy Storage Capacity (MWh)	Power (MW)	Duration (h)
Hornsdale Power Reserve	Australia	129	30+70	3
Escondido Substation	U.S.	120	30	4
Pomona Substation	U.S.	80	20	4
Mira Loma Substation	U.S.	80	20	4
Tesla Solar Plant	U.S.	52	13	5
Stocking Pelham facility	England	50	50	1
Jardelund	Germany	50	48	1
Minamisōma Substation	Japan	40	40	1

Figure 1.33: Artist rendition of Li-ion battery storage facility showing grid storage for renewable solar and wind energy resources; https://www.shutterstock.com/image-illustration/battery-energy-storage-concept-nice-morning-765921109?src=jfSS5w5GD7eikq1CJl-SNA-1-3.

Vanadium flow batteries, as described above, have also played an important role in grid storage. Table 1.9 summarizes the world's largest vanadium flow battery facilities. An 800 MWh, 200 MW vanadium flow battery storage facility is under construction in Liaoning, Dailan, China and is expected to be operational by around 2020. A photograph of a vanadium flow battery for grid storage is shown in Figure 1.34. As is often the case for grid storage batteries of all chemistries, this battery is fabricated inside a shipping container to facilitate transportation and installation.

Numerous smaller battery storage facilities exist worldwide using the three battery technologies above, as well as other types of batteries, such as lead-acid and nickel-cadmium.

Table 1.9: World's largest vanadium flow battery grid storage facilities

Name	Country	Energy Storage Capacity (MWh)	Power (MW)	Duration (h)
Minami Hayakita Substation	Japan	60	15	4
Woniushi, Liaoning	China	10	5	2
Zhangbei Project	China	8	2	4
SnoPUD MESA 2 Project	U.S.	8	2	4
San Miguel Substation	U.S.	8	2	4
Tomamae Wind Farm	Japan	6	4	1.5

Figure 1.34: Vanadium flow battery constructed in a shipping containe; https://www.shutter-stock.com/image-photo/cornwall-united-kingdom-november-2017-energy-1241178277?s-rc=CMkTM-HSzuq0JO9smuqkiA-1-2.

While there are a significant number of grid storage facilities that utilize batteries of different designs, the largest grid storage facilities use pumped hydroelectric technology and can provide up to 2–3 GW of power. However, battery storage for the grid can effectively complement large-scale pumped hydroelectric storage. Batteries are appropriate for short-term grid stabilization and in many cases may be the most appropriate for integration with renewable energy sources because of the relatively undemanding geographical requirements. Integration of battery storage with either photovoltaic or wind energy resources follows along the lines of the diagrams in Figures 1.29 and 1.30, respectively, except that a synchronizing breaker is used to connect the AC output to the grid in order to synchronize the phase and frequency of the AC currents.

CHAPTER 2

Supercapacitors and Superconductors

2.1 INTRODUCTION

Energy may be stored in an electric or magnetic field. In the former case, electricity is used to create a charge distribution that produces the electric field that stores the energy. The simplest device that can be used for this purpose is the capacitor. When the plates of the capacitor are charged by an electric current, an electric field is produced which stores energy. Energy is recovered when the capacitor is discharged to produce an electric current. Supercapacitors store energy both in an electric field and electrochemically. In the case of energy storage in a magnetic field, an electric current flowing through a coil of wire produces the magnetic field. In order to avoid resistive losses in the coil, superconducting materials are used to carry the current. Energy is recovered by extracting the current from the coil. These two approaches to storing energy in electric and magnetic fields are described in the present chapter.

2.2 CAPACITORS AND SUPERCAPACITORS

2.2.1 ENERGY STORAGE IN A CAPACITOR

Consider a parallel plate capacitor, as shown in Figure 2.1, with a distance d between the plates and an area of each plate of A. The capacitor is charged by moving charges (say negative charges) from one plate (which will be left with a positive charge) to the other plate (which accumulates the negative charge). If the charge on each plate is Q, then Gauss' law gives the electric field, $\mathcal{E}$, between the plates as

$$\mathcal{E} \frac{Q}{\varepsilon_r \varepsilon_0 A}, \tag{2.1}$$

where ε_0 is the permittivity of free space (8.85×10^{-12} F/m; F = farad = J/V^2) and ε_r is the relative permittivity of the dielectric medium between the plates. The voltage across the capacitor is determined by integrating the electric field from one plate to the other or

$$V = \int_0^d \mathcal{E}dx = \mathcal{E}d = \frac{Qd}{\varepsilon_r \varepsilon_0 A}. \tag{2.2}$$

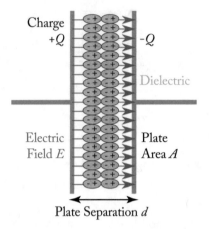

Figure 2.1: **A charged parallel plate capacitor;** https://commons.wikimedia.org/wiki/File:Capacitor_schematic_with_dielectric.svg, https://creativecommons.org/licenses/by-sa/3.0/deed.en.

Capacitance is defined as the ability to store charge:

$$C = \frac{Q}{V} .$$
(2.3)

So, for the parallel plate capacitor, this is

$$C = \frac{\varepsilon_r \varepsilon_0 A}{d} .$$
(2.4)

When the capacitor is charged, energy is stored in the electric field between the plates. We can calculate the stored energy in the following way. Consider a charged capacitor with a voltage V between the plates. If we move an additional differential charge, dq, from one plate to the other, the work (i.e., energy) required to do this, dE, is

$$dE = Vdq.$$
(2.5)

Substituting for the voltage from Equation (2.3) gives

$$dE = \frac{qdq}{C} .$$
(2.6)

Thus, to charge the capacitor to a total charge Q requires

$$E = \int dE = \int_0^Q \frac{q}{C} dq = \frac{1}{2} \frac{Q^2}{C} .$$
(2.7)

This is, therefore, the energy stored in the capacitor when the charge on the plates is Q. Since it is easier to directly measure voltage rather than charge, we can use Equation (2.3) to give

$$E = \frac{1}{2}CV^2. \tag{2.8}$$

From Equations (2.4) and (2.8) we can calculate the energy storage capacity of a parallel plate capacitor as

$$E = \frac{\varepsilon_r \varepsilon_0 A V^2}{2d}. \tag{2.9}$$

While parallel plate capacitors are not a practical means of storing energy, Equation (2.9) provides us with some insight into the design parameters that are of relevance in designing a practical capacitor. Energy storage is optimized by maximizing ε_r, A, and V, while minimizing d. ε_r is obviously a material dependent quantity and will influence our choice of the dielectric material between the capacitor electrodes. A and d depend on geometry of the capacitor. All of these factors will be discussed further below. However, we need first to consider the effects of voltage.

Increasing the voltage between the capacitor plates will increase its energy storage capacity. However, this cannot be done indefinitely. At some point the dielectric will start conducting and the capacitor will discharge. This point occurs when the electric field in the dielectric reaches its breakdown value, ε_b. The breakdown voltage, V_b, is related to this field by

$$V_b = d\mathcal{E}b. \tag{2.10}$$

The maximum energy storage capacity is given by substituting Equation (2.10) into Equation (2.9) to give

$$E_{max} = \frac{1}{2} A\varepsilon_r \varepsilon_0 \, \mathcal{E}b^2, \tag{2.11}$$

where the breakdown electric field is an intrinsic property of the dielectric material. Although one can always increase the energy storage capacity by increasing the size of the capacitor, it is most useful to consider the energy storage capacity per unit mass. Since the volume of the parallel plate capacitor is Ad, then its mass will be $Ad\rho$, where ρ is the average density. Thus, Equation (2.11) may be rewritten as

$$\left(\frac{E}{m}\right)_{max} = \frac{1}{2}\frac{\varepsilon_r \varepsilon_0}{\rho} \, \mathcal{E}b^2. \tag{2.12}$$

Capacitors have a wide variety of uses in electronic circuits. Among these are applications in timing circuits and filters. For many such applications, traditional capacitors are appropriate. These fall into several categories depending on their construction. Figure 2.2 shows the relationship of the major types of capacitors. Capacitors are either non-polarized or polarized. The non-polarized capacitors have electrodes that are symmetric (as suggested in Figure 2.1), and the positive voltage can be applied to either electrode. Polarized capacitors have electrodes that are not symmetric and require that the positive voltage be applied to only one of the electrodes. Electrolytic capacitors are a common type of polarized capacitor and fall into three categories, aluminum, tantalum and niobium, depending on the electrode material. Figure 2.3 shows the design of a typical electrolytic capacitor. The two electrodes are distinguished by their surface treatment. The anode (positive electrode) is etched to oxidize the surface and create an insulating layer.

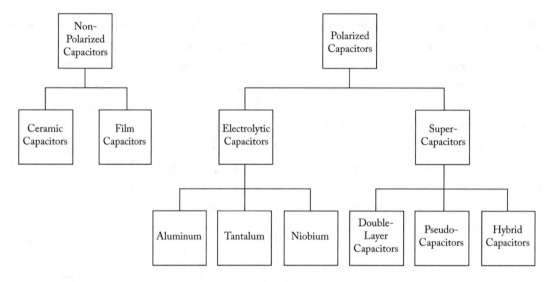

Figure 2.2: Categories of capacitors.

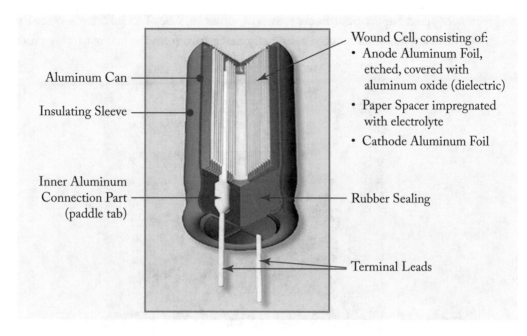

Aluminum Can

Insulating Sleeve

Inner Aluminum
Connection Part
(paddle tab)

Wound Cell, consisting of:
• Anode Aluminum Foil,
  etched, covered with
  aluminum oxide (dielectric)
• Paper Spacer impregnated
  with electrolyte
• Cathode Aluminum Foil

Rubber Sealing

Terminal Leads

Figure 2.3: Construction of a typical aluminum electrolytic capacitor; https://upload.wikimedia.org/wikipedia/commons/f/f3/Al-e-cap-construction.jpg.

2.2.2 SUPERCAPACITORS

Supercapacitors (also called ultracapacitors) are a class of polarized capacitors that has an energy storage capacity per unit mass that is a factor of 10–100 times that of traditional capacitors; see Figure 2.4. Supercapacitors are, therefore, the most suitable type of capacitor for storing energy. As Figure 2.2 shows, they may be categorized as double-layer capacitors, pseudocapacitors, or hybrid capacitors. We begin with the simplest of these, the double-layer capacitor.

Figure 2.5 shows a diagram of a simple double-layer capacitor. The device consists of two electrodes separated by a space containing an electrolyte and a separator, as shown in the figure. The electrolyte contains both positive and negative ions and the separator is an ion-permeable membrane. In the uncharged state (shown in Figure 2.5(a)), the positive and negative ions are randomly distributed throughout the electrolyte. When a voltage is applied to the supercapacitor, as in Figure 2.5(b), the ions in the electrolyte interact with the electric field present between the electrodes. They pass through the permeable membrane and are redistributed as shown. Thus, the device looks like two charged capacitors in series with negatively and positively charged layers formed at each electrode and this results in the polarization of a layer of the dielectric electrolyte molecules known as the Helmholtz plane. Supercapacitors are polarized, either because of an inherent asymmetry in the electrodes or as a result of an induced asymmetry that results from the manufacturing process.

Figure 2.4: A 500 farad supercapacitor. The capacitor is about 3.5 cm in diameter and 6.0 cm long. A U.S. 25¢ coin is shown for comparison (image courtesy of Richard A. Dunlap).

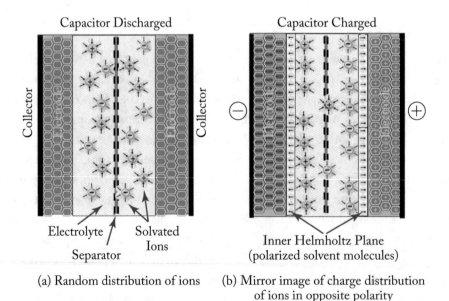

(a) Random distribution of ions

(b) Mirror image of charge distribution of ions in opposite polarity

Figure 2.5: Supercapacitor (a) in the uncharged state and (b) in the charged state (right); https://commons.wikimedia.org/wiki/File:EDLC-Charge-Distribution.png.

The total capacitance of the double-layer supercapacitor is the net capacitance of the two capacitors in series or

$$C = \frac{C_1 C_2}{C_1 + C_2}.$$

(2.13)

The relative values of C_1 and C_2 depend on the design of the capacitor.

The origin of the capacitance of the supercapacitor is two-fold: the normal electrostatic capacitance, as described above, and electrochemical pseudocapacitance, as illustrated in Figure 2.6, due to the adsorption of ions on the electrode surface.

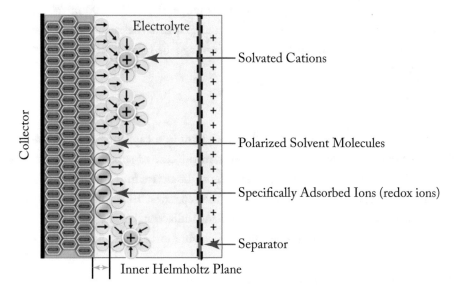

Figure 2.6: Description of electrochemical pseudocapacitance of a double-layer supercapacitor due to the adsorption of ions on the electrode; https://commons.wikimedia.org/wiki/File:Pseudocapacitance-Priciple.png.

The relative importance of the electrostatic capacitance and the pseudocapacitance depends on the specific design of the capacitor. For double-layer supercapacitors, the normal electrostatic capacitance dominates over the electrochemical pseudocapacitance.

The optimization of the capacitive properties of the supercapacitor comes, to a large extent, from the proper selection of the materials used in its construction. Here we specifically consider the suitability of materials for the electrodes, electrolyte, and the separator.

Electrodes

Equation (2.9) shows that the amount of energy stored in a capacitor is proportional to the ratio *A/d*. An inspection of Figure 2.5 shows that the distance between the positive and negative charges near each electrode is of the order of the thickness of the Helmholtz plane of dielectric molecules. This plane is typically in the range of 0.3–0.8 nm in thickness, much less than the dielectric thickness in a normal capacitor. Thus, the supercapacitor has an inherently high capacitance because of the small value to d in Equation (2.9). However, the capacitance is further enhanced by the choice of an electrode material that provides a very large surface area, A. Supercapacitors typically use activated carbon as an electrode material. Activated carbon is very porous and, as a result, has enormous surface area per unit mass, typically in the range of 1,000–3,500 m^2/g. Nanostructured carbons, such as carbon fibers, carbon nanotubes, and graphene-based materials, also offer desirable surface to mass ratios but are generally less advantageous from an economic standpoint than traditional activated carbon.

Electrolyte

From Equation (2.12) The effectiveness of the electrolyte in a capacitor depends on its dielectric constant and its breakdown electric field. However, in the case of supercapacitors, there are other criteria that need to be considered. The electrolyte must dissociate into an equal number of positive cations and negative anions that can pass through the separator. The electrolyte must also provide the dielectric molecules that form the Helmholtz planes adjacent to the positive and negative electrodes. Since the electrodes are very porous, the electrolyte must be a liquid so that it can penetrate the pores.

Two types of electrolytes are commonly used for double-layer supercapacitors: aqueous and organic. Aqueous electrolytes are water-based solutions containing inorganic chemicals. Inorganic additives may include sulfuric acid, potassium hydroxide, sodium perchlorate, or lithium perchlorate. Supercapacitors with aqueous electrolytes tend to have low specific energy but high specific power. Organic solvents, such as acetonitrile or tetrahydrofuran, may also be used as electrolytes. Additions to these organic solvents to provide ions include ammonium salts or alkyl ammonium salts. Supercapacitors with organic electrolytes tend to have high specific energy and low specific power. The relationship between specific energy and specific power is illustrated in Figure 2.7. It is seen in the figure that supercapacitors (EDLC) have higher specific power but lower specific energy than batteries and that there is a trade-off between energy and power.

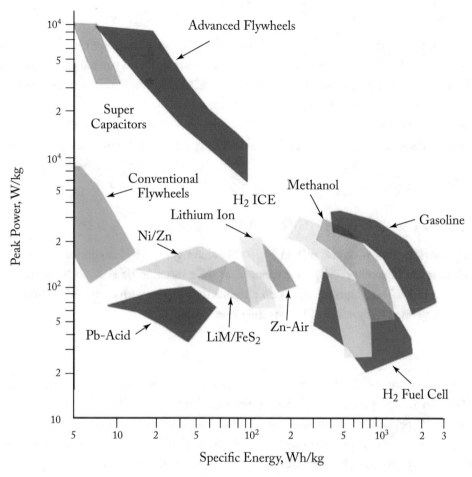

Figure 2.7: Relationship of specific energy and specific power (Ragone plot) for some energy storage devices. Based on Ghoniem (2011), reprinted with permission from Elsevier.

Separator

The separator is designed to electrically separate the two electrodes. It must be thin or porous enough to allow the conducting ions to pass through without unnecessarily adding internal resistance to the capacitor and thereby contributing to self discharge. Typically insulating material with a thickness of a few tens of micrometers is used. Suitable materials include paper (for inexpensive devices), organic films such as polyacrylonitrile or Kapton, or woven glass or ceramic fibers.

2.2.3   PSEUDOCAPACITORS AND HYBRID CAPACITORS

Pseudocapacitors are similar in design to double-layer supercapacitors except that they are designed so that the electrochemical pseudocapacitance dominates over the conventional electrostatic capacitance. These devices typically increase the electrochemical pseudocapacitance by using metal oxides, such as MnO_2 or RuO_2, as electrode materials rather than carbon.

Hybrid capacitors are intermediate between double-layer supercapacitors and pseudocapacitors. They are designed with two distinctly different electrodes, one dominated by the electrostatic capacitance of the Helmholtz layer and the other dominated by the electrochemical pseudocapacitance.

2.2.4   APPLICATIONS OF SUPERCAPACITORS

Supercapacitors, as shown on the plot in Figure 2.7, have very different characteristics than batteries. Their much higher specific power means that their recharge time is much shorter than rechargeable batteries. Thus, supercapacitors do not directly compete with batteries but offer different characteristic that are suitable for different situations.

Personal Electronics

While supercapacitors have a lower specific energy capacity than batteries, they have been used successfully for the purpose of stabilizing power supplied to portable devices such as notebook computers, GPS devices, cell phones, etc. They are also useful for many such devices as short-term back-up power and/or emergency shutdown power. Because of their significant specific power, supercapacitors have proved to be useful for devices that require short periods of high power such as photographic flash lamps. The more rapid recharge time, compared to rechargeable batteries, also benefits this application.

Power Buffer

Along similar lines to stabilizing power for personal battery-operated devices but on a larger scale, supercapacitors have been found to be useful to buffer grid fluctuations caused by short-term high current loads. A common example of such a situation is during the charging of batteries for battery electric vehicles or plug-in hybrids. Appropriate supercapacitor buffering may be integrated into charger electronics to prevent the charger load from introducing fluctuations in the grid supply.

Vehicle Applications

Energy storage capacity and how it affects vehicle range is a major concern for battery electric vehicles. Because of a lower specific energy capacity, supercapacitors are not appropriate as a sole means of energy storage for electric vehicles. However, their high specific power provides a suitable means of supplementing power in a battery electric vehicle or hybrid vehicle. This an attractive approach for racing cars, as the large specific power of supercapacitors makes them preferable to batteries for hybrids that require rapid recharging from regenerative braking and short periods of quickly available power. Figure 2.8 shows a prototype Toyota TS030 supercapacitor hybrid. Vehicles that use this approach have been quite successful against flywheel hybrid competitors such as the Audi R18 e-tron quattro.

Figure 2.8: The Toyota TS030 Hybrid, a Le Mans Prototype 1 (LMP1) racing car which uses supercapacitors; https://commons.wikimedia.org/wiki/File:Le_Mans_2013_(9347549094),jpg, https://creativecommons.org/licenses/by/2.0/deed.en.

The fast charging and high-power output features that make supercapacitors beneficial in racing cars may also make them appropriate as a supplementary power system for passenger vehicles. Many automobile manufacturers have developed prototype or experimental vehicles that use supercapacitors to best take advantage of energy recovery through regenerative braking in conjunction with batteries in either battery electric vehicles or gasoline-electric hybrids. It is probable that further technical developments will result in the commercial availability of such vehicles. The

Mazda 6 is a gasoline powered vehicle (Figure 2.9) that uses a unique supercapacitor-based system called *i-eloop*. Energy recovered from regenerative braking is used to charge a supercapacitor. This energy is then used to power the vehicles electrical accessories, thereby eliminating the need to use engine power for purposes other than vehicle propulsion. Mazda claims that this approach reduces fuel consumption by about 10%.

Figure 2.9: The 2018 Mazda 6, which uses supercapacitors rather than batteries for energy storage; Image from Grzegorz Czapski/Shutterstock.com, https://www.shutterstock.com/image-photo/gene-va-switzerland-march-06-2018-metallic-1058170427?src=73j44uMUS6nGt3c_Almfhw-1-0.

In the past decade, a number of prototype systems have been installed, mostly in Europe and China, that use supercapacitors to power light-rail vehicles (sometimes called trams). These vehicles traditionally utilize power obtained from overhead catenary power lines. While the energy stored in supercapacitors can only power such a vehicle for a few kilometers, at most, it allows for a transportation system where overhead power lines do not need to be contiguous. This has several potential advantages.

- It eliminates the need for overhead power lines in city locations where their installation is expensive, difficult, or even impossible.

- It eliminates the need for overhead power lines in locations where they could distract from a city's architectural heritage.

- It can reduce overall energy consumption.

- It can reduce peak demand on the electric grid that can occur at rush hour.

The use of supercapacitors rather than batteries has the advantage of short recharging times which can allow the supercapacitors to recharge from overhead or ground-level connections during stops.

2.3   SUPERCONDUCTING MAGNETIC ENERGY STORAGE

2.3.1   ENERGY STORAGE CAPACITY OF AN INDUCTOR

A superconducting magnet consists of a coil of superconducting wire. In order to determine the energy storage capabilities of a superconducting coil, we begin with an analysis of a simple coil in an external magnetic field. Faraday's law states the following.

The emf induced in a circuit is proportional to the time rate of change of the magnetic flux through any surface that is bounded by the circuit.

This law may be expressed mathematically as

$$\mathcal{E} = -\frac{d\phi}{dt}, \qquad (2.14)$$

where $\mathcal{E}$ is the induced emf and ϕ is the magnetic flux. The negative sign on the right-hand side of Equation (2.13) comes from Lenz's law which states that the induced emf will produce a magnetic field that opposes the change in the external field. This same phenomenon results from changes in a current in the circuit itself. A current flowing though a coil produces a magnetic field. If the current changes in time, then Faraday's law states that there will be a proportional emf induced in the circuit. In this case, the phenomenon is referred to as self-inductance. As a simple example, let us consider a solenoid of length l consisting of N turn of wire and a cross sectional area A carrying a current I. Ampere's law can be easily applied to find the magnetic field inside the solenoid as

$$B = \frac{\mu_0 NI}{l}, \qquad (2.15)$$

where μ_0 is the permeability of free space ($\mu_0 = 1.26 \times 10^{-6}$ N/A^2). The total magnetic flux through N coils is related to the magnetic field as

$$\phi = NBA, \qquad (2.16)$$

and the above equations can be combined to give

$$\mathcal{E} = -\frac{\mu_0 N^2 A}{l}\frac{dI}{dt} = -\mu_0 n^2 Al\frac{dI}{dt}, \qquad (2.17)$$

where $n = N/l$ is the number of turns per unit length. This is often written as

$$\mathcal{E} = -L\frac{dI}{dt},$$

(2.18)

where

$$L = \mu_0 n^2 A l$$

(2.19)

is defined as the self-inductance (usually just called the inductance) of the solenoid. Expressions for the inductance of other coil geometries may also be determined.

The inductor can store energy associated with the magnetic field it produces when a current is passing through the coils. To calculate the energy storage capacity of an inductor we look at a simple LR circuit, as shown in Figure 2.10, where a battery provides an emf of $\mathcal{E}_0$. When the switch is closed the emf around the loop must add to zero or

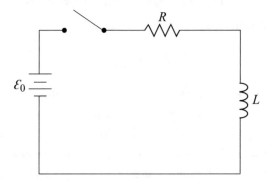

Figure 2.10: Simple LR circuit with a battery.

$$\mathcal{E}_0 - IR + \mathcal{E} = 0,$$

(2.20)

where IR is the voltage drop across the resistor and $\mathcal{E}$ is the emf in the inductor as above. Multiplying Equation (2.19) by the current gives

$$\mathcal{E}_0 I - I^2 R + \mathcal{E}I = 0,$$

(2.21)

where the three terms on the left-hand side of the equation can be interpreted as the power provided by the battery, the power dissipated by the resistor, and the power stored by the inductor, P, respectively. This last term may be written in terms of Equation (2.17) as

$$P = LI\frac{dI}{dt}.$$

(2.22)

The energy stored in the inductor can be found by integrating Equation (2.21) as

$$E = \int P \, dt = \int_0^I LI \, dI = \frac{1}{2}LI^2 \, . \qquad (2.23)$$

This gives the total energy stored when the current in the inductor has a value I. Thus, the energy storage capacity of an inductor may be determined as a function of the current flowing through the inductor and the value of the inductance as determined for the particular coil geometry. While Equation (2.18) is appropriate for calculating the inductance of a long solenoid, actual coil geometries are typically not long solenoids. One approach, which is suitable for devices which are not too large, is a short solenoid. In this case, Equation (2.18) for the inductance may be modified to

$$L = RN^2 f \, , \qquad (2.24)$$

where R is the mean radius, N is the total number of turns, and f is a geometry dependent factor, referred to as the form function. This gives a total energy storage capacity of

$$E = \frac{1}{2} RN^2 f I^2 \, . \qquad (2.25)$$

For large devices a toroidal coil geometry may also be suitable. It is important when using Equation (2.22) or (2.24) to ensure that proper units are used. In the SI system inductance is measured in Henrys (H). The Henry can be expressed in fundamental SI units as

$$H = \frac{kg \cdot m^2}{s^2 \cdot A^2} = \frac{J}{A^2} \, , \qquad (2.26)$$

where A is amperes and J is joules.

2.3.2   SUPERCONDUCTIVITY

An inductor can be used to store energy in its magnetic field as described above. However, inductors made of normal conducting material are not pure inductors because the wire has resistance as well. This means that once the coil is disconnected from the power source, then the energy will dissipate through the resistance of the inductor. For this reason, energy storage devices which store energy in the form of a magnetic field utilize superconducting coils.

Superconductivity was discovered in 1911 by Heike Kamerlingh Onnes, who observed that, when mercury was cooled to a temperature below about 4.1 K, the electrical resistivity dropped to a value that was too low to be measured experimentally. This property is known as superconductivity and, subsequent to Kamerlingh Onnes' discovery, has been observed in a large number of elements, alloy, and compounds. These materials have a number of very significant applications that include electric power transmission, energy storage, and sensors. However, properties of superconductors vary greatly from one material to another and an analysis of their potential uses requires a detailed consideration of these properties.

Each superconducting material has a critical temperature, T_c, below which the material becomes superconducting and above which the material possesses normal resistive electrical behavior. Merely cooling the material to a temperature below T_c is not necessarily sufficient to make use of its superconducting properties. Superconductivity is closely related to magnetism and the application of a sufficiently large magnetic field destroys the superconducting properties. The critical field is related to the critical temperature by the relationship

$$B_c(T) = B_c(0)\left[1 - \left(\frac{T}{T_c}\right)^2\right],$$
(2.27)

where $B_c(0)$ is the critical field at zero temperature. It is clear that, as the temperature approaches the critical temperature from below, the critical field goes to zero. Thus, at a given temperature below T_c, as the magnetic field is increased, a point is reached at which the material suddenly loses its superconductivity and becomes normal.

Superconductors which follow the relationship given by Equation (2.26) are referred to as type-I superconductors. While some elements exhibit type-I superconductivity, most superconducting materials exhibit what is referred to as type-II superconductivity. Figure 2.11 shows the difference between a type-I and a type-II superconductor in terms of the temperature dependence of the critical field. If a type-II superconductor at a temperature below the critical temperature is subject to an increasing magnetic field, then at some point the material will enter into a mixed state where part of the superconductor becomes normal and the rest of the material remains superconducting. Since an electric current follows the path of least resistance, the current flows without resistance through the part of the sample that remains superconducting.

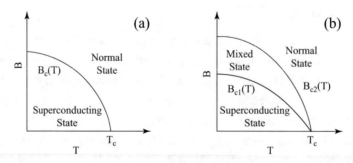

Figure 2.11: Phase diagram of (a) type-I superconductor and (b) type-II superconductor.

A comparison of the properties of some superconductors is given in Table 2.1. It is clear that, from a practical standpoint, type-II superconductors show much superior critical temperatures and critical fields, thereby making them the obvious choice for most superconductor applications.

Table 2.1: Properties of some superconducting materials

Material	Type	T_c (K)	$B_c(0)$ (T)	$B_{c2}(0)$
Sn	I	3.7	0.03	-
Pd	I	7.2	0.08	-
Nb	II	9.3	-	0.4
NbTi	II	10	-	15
Nb3Sn	II	18.3	-	30
Nb3Ge	II	23.2	-	37

Since a current flowing through a wire produces a magnetic field, there is a critical current, Ic, at which the superconductivity is destroyed, that follows the same trend as the critical field. It is, therefore, desirable to use superconductors at a temperature as far below T_c as possible to best make use of their properties.

Since the discovery of superconductivity, the search for materials with the highest possible critical temperature and critical field has been a major component of superconductivity research. Figure 2.12 shows the maximum observed value of T_c as a function of year. From the late 1940s to the mid-1980s, Nb_3Ge remained the highest T_c superconductor. In 1986 the first of the so-called high temperature superconductor, lanthanum barium copper oxide (LaBaCuO, also known as LBCO), was discovered by J. Georg Bednorz and K. Alex Müller. Not only was the critical temperature 12 K higher than that previously reported for Nb_3Ge, but it was the first instance where a material that was an insulator at room temperature became a superconductor at low temperature. This discovery led to further research that raised the maximum value of T_c to around 150 K by 1995. Recently, H2S, which is a gas at standard pressure and temperature (STP), was found to become superconducting at around 200 K under high pressure. While this material has not found any practical applications at present, it has led to a new approach to thinking about superconductors with high critical temperatures. In fact, there has been speculation that some basic hydrogen compounds may, under sufficient pressure, become superconducting near room temperature.

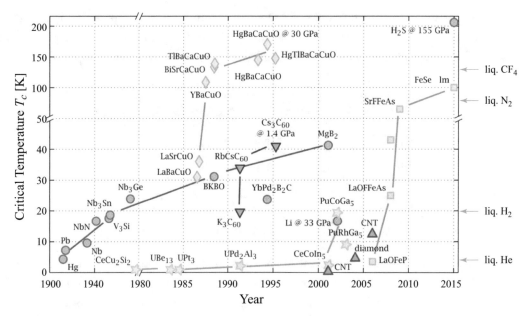

Figure 2.12: Maximum experimentally measured values of T_c for superconductors as a function of year since their first discovery in 1911. Based on Ray (2015). Image available at https://commons.wiki-media.org/wiki/File:Timeline_of_Superconductivity_from_1900_to_2015.svg.

Table 2.2 gives some properties of two of the most common high temperature superconductors shown in Figure 2.12. High-temperature superconductors are all type-II superconductors, and the table shows that they not only have high critical temperatures, but also exhibit large values of the critical field.

Table 2.2: Critical temperatures and fields for some high-temperature superconductors

Material	Common Designation	T_c (K)	B_{c2} (K)
$YBa_2Cu_3O_7$	YBCO	93	168
$Bi_2Sr_2Ca_2Cu_3O_{10}$	BSCCO	110	200

2.3.3 SUPERCONDUCTING MAGNETIC ENERGY STORAGE

One of the potential applications of superconducting materials is the storage of substantial amounts of energy in the large magnetic field which can be produced. The temperature and magnetic field produced by the electric current must, of course, be appropriate for the material to remain in the superconducting regime.

A superconducting magnetic energy storage system consists of three principal components:the superconducting coil, a cryogenic refrigeration system, and a control system for injecting and extracting the current. For devices that are intended for integration into the grid, the electrical energy to be injected or extracted from the grid is in the form of AC, while the superconducting magnet stores DC. Therefore, the control system must convert AC to DC for storage using a rectifier and DC to AC for energy recovery using an inverter. While the superconducting coil stores electrical energy without loss, the rectifier/inverter has a typical net efficiency of about 95%. However, there is an energy cost for superconducting magnetic energy storage because energy is required to maintain the low temperature necessary to keep the magnet in the superconducting state.

The choice between conventional low temperature superconducting materials and high-temperature superconducting materials is an important consideration for the construction of a superconducting magnetic energy storage device. While it would seem that the use of high-temperature superconductors would simplify the cooling requirements for the magnet, there are other factors that are important as well. The manufacture of conventional superconducting wire, e.g., NbTi, is a relatively straightforward matter. High-temperature superconductors are ceramics and are, therefore, brittle. The preparation of a suitable conductor from which a coil can be manufactured is not so straightforward. The most suitable method of dealing with this difficulty is the preparation of superconducting tapes, as shown in Figure 2.13. A micron-thick layer of superconducting ceramic is deposited onto a substrate and integrated into a supporting structure. The superconducting layer is less than 1% of the total thickness of the tape. The preparation of such a tape is expensive and increased infrastructure costs must be weighed against potential savings in operational costs. It turns out that refrigeration costs are a minor component of overall cost and become less important for larger scale devices. It is also important to note that high-temperature superconductors benefit from a greater critical field by being used at lower temperatures, so one may choose to operate at temperatures that are close to those used for low-temperature superconductors, in any case. The net outcome of these considerations is that, at present anyway, the benefits of high-temperature superconductors are not so obvious. As a result, commercial superconducting magnetic energy storage devices are manufactured using conventional superconducting materials, typically NbTi or Nb_3Sn. Further research and technological developments may change this in the future.

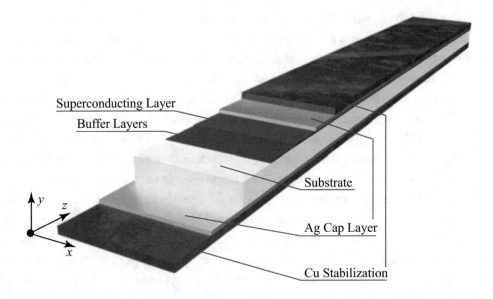

Figure 2.13: Construction of a high temperate cuprate superconductor power transmission cable. Based on Solovyov et al. (2013), with permission of IOP Publishing.

There are three potential applications for superconducting magnetic energy storage devices:

- grid backup,

- grid stabilization, and

- pulsed power.

These are discussed next.

Grid Backup

There is the possibility that superconducting magnetic energy storage could fill the same needs as pumped hydroelectric storage or large-scale battery energy storage. While no device has been constructed that could satisfy these requirements, studies have considered the parameters of a suitable system. Table 2.3 provides some measurements and operating conditions that would be appropriate for such a device. As the table shows, the device would be 1 km in diameter. Cooling the magnet would require a vacuum enclosure of this diameter to surround the liquid helium cryogen. Such a device is a major technical challenge and will require the accumulation of knowledge based on the development of smaller-scale superconducting magnetic energy storage devices as detailed below.

Table 2.3: Characteristics of a superconducting magnetic energy storage plant that would satisfy the needs now met by pumped hydroelectric storage. Information adapted from P. Tixador (January, 2008)

Quantity	Value	Units
Superconductor	NbTi	-
Energy	5250	MWh
Power	1000	MW
Magnet diameter	1000	m
Magnet height	19	m
Current	200	kA
Temperature	1.8	K

Grid Stabilization

While superconducting magnet grid-scale energy backup is in the very early experimental and development stages, grid stabilization is an established and commercialized application of superconducting magnetic energy storage. These devices, which typically utilize NbTi superconducting wire, are used as a means of reducing fluctuations in grid voltage. These are useful for the purpose of producing high-quality stabilized grid voltages for research or manufacturing facilities that are sensitive to voltage fluctuations. They are also useful for stabilizing voltage fluctuations that could lead to instability in the grid and possible power outages. Superconducting magnetic energy storage devices are in fairly common use by power companies for the purpose of grid stabilization and a typical device is illustrated in Figure 2.14.

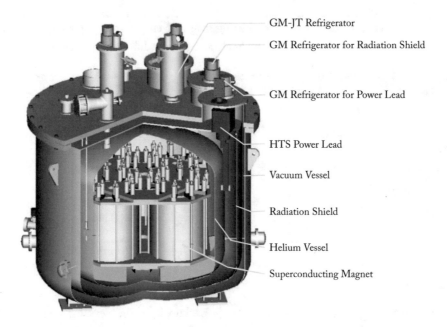

GM-JT Refrigerator

GM Refrigerator for Radiation Shield

GM Refrigerator for Power Lead

HTS Power Lead

Vacuum Vessel

Radiation Shield

Helium Vessel

Superconducting Magnet

Figure 2.14: Superconducting magnetic energy storage device used to stabilize grid fluctuations. This device can produce up to 10 MW output for up to 3 s and has a total energy storage capacity of about 8 kWh. Reproduced and based on Nagaya et al. (2006) with permission from IEEE.

Pulsed Power

Some electrical applications require large amounts of power over fairly short periods of time and superconducting magnetic energy storage devices are ideal for this purpose. One such application is the electromagnetic launcher, based on the design commonly known as a railgun; see Figure 2.15. When the switch is closed, a current loop is formed through the rails and projectile. This current produces a magnetic field perpendicular to the plane of the rails. There is a Lorentz force on the charge carriers in the conductors. Since the rails are fixed, they cannot move, but the projectile, which can slide on the rails is propelled away from the current supply. These devices can launch objects at velocities of up to 2 km/s.

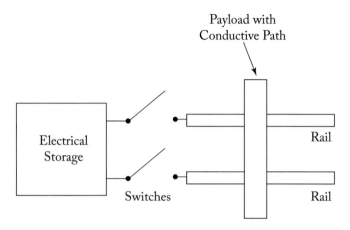

Figure 2.15: Schematic design of a railgun.

There are several potential future applications for rail guns including, on the small-scale end, weapons to shoot projectiles at high velocity without the need for explosives. Prototype devices have been developed and are being tested. On the large-scale end of applications, these devices have been proposed as a means of launching spacecraft or for launching aircraft from limited spaces such as aircraft carriers.

CHAPTER 3

Chemical Energy Storage Methods

3.1 INTRODUCTION

The energy storage methods described thus far, have dealt with the storage of electricity. This is the most appropriate means of energy storage for the purpose of grid stabilization, grid backup, and residential electricity. While electric vehicles have become increasingly popular, as illustrated in Chapter 1, the use of chemical energy storage methods may be an attractive alternative for transportation. Energy produced by renewable techniques, which is in the form of electricity, may be stored by using this energy to produce chemical compounds from which this energy can later be recovered by combustion or other means. Several possible compounds have been considered for chemical energy storage. We begin with the simplest of these: molecular hydrogen.

3.2 HYDROGEN

While hydrogen is sometimes promoted as an energy source for the future, it is, in fact, not a source of energy, but merely a means of storing energy. There are no sources of pure hydrogen in nature, as there are other sources of energy that can be harvested, such as petroleum and coal that can be extracted from the earth, wind, or solar energy that can be harvested from nature using turbines or photovoltaics. Hydrogen must be produced artificially by extracting it from hydrogen-containing compounds using energy obtained from energy sources in nature. The simplest of these hydrogen production methods, and perhaps the best known, is the production of molecular hydrogen gas, H_2, by the electrolysis of water. In the present section we overview some of the relevant properties of hydrogen, as well as the ways in which it can be produced and subsequently used to power vehicles or produce electricity.

3.2.1 PROPERTIES OF HYDROGEN

Hydrogen is the lightest element. There are two stable isotopes of hydrogen, ^1H with nuclei containing a single proton and ^2H with nuclei containing one proton and one neutron. Natural hydrogen consists of 99.985% ^1H and 0.015% ^2H. Table 3.1 summarizes some of the basic physical and chemical properties of hydrogen. The energy content of hydrogen (that is the energy that is stored in hydrogen, which can be recovered) is given by the higher heat of combustion in the table. This energy becomes available when hydrogen is burned according to the reaction

$$2H_2 + O_2 \rightarrow 2H_2O. \tag{3.1}$$

When the hydrogen is burned, the by-product (water) is in the gaseous state. The lower heat of combustion is the energy which becomes available immediately. The higher heat of combustion is the total energy which becomes available when the by-product is returned to room temperature and includes, most importantly, the latent heat of vaporization of the water.

Table 3.1: Properties of hydrogen, H_2. Note: standard temperature and pressure (STP) = 293 K, 10^5 Pa

Quantity	Value	Units
Molecular weight	2.01588	g/mol
Boiling point (10^5 Pa)	20.27	K
Melting point	13.99	K
Gas density (STP)	0.08988	kg/m^3
Liquid density (20.27 K)	70.99	kg/m^3
Lower heat of combustion	119.96	MJ/kg
Lower heat of combustion (H2 gas)	10.78	MJ/m^3
Lower heat of combustion (H2 liquid)	8516	MJ/m^3
Higher heat of combustion	141.88	MJ/kg
Higher heat of combustion (H2 gas)	12.7	MJ/m^3
Higher heat of combustion (H2 liquid)	10,072	$MJ/m3^3$

The energy content of hydrogen can be compared with that of some other common fuels as shown in Table 3.2. It is clear that the energy content, ~142 MJ/kg, is higher than that of the other fuels in the table. Most notably, the energy content of hydrogen per unit mass is more than three times that of gasoline. As the table shows, however, the energy content per unit volume is much lower than that of gasoline (or any of the other fuels that are liquid at STP). Even in the liquid state, the volumetric energy density of hydrogen is lower than other fuels because of the low density of hydrogen liquid. This feature is clearly seen by plotting the energy per unit volume as a function of energy per unit mass, as shown in Figure 3.1. This property of hydrogen emphasizes one of the difficulties in using hydrogen as a fuel, particularly for applications such as transportation, which require portability.

Table 3.2: Comparison of the energy content of some common fuels. The energy is the higher heat of combustion as defined for Table 3.1

Material	Formula	Molecular Weight (g/mol)	Density (STP) (kg/m³)	Energy (MJ/kg)	Energy (MJ/m³)
Hydrogen	H_2	2.01588	0.08988	141.88	12.7
Methane	CH_4	16.04	0.656	55.5	36.4
Ammonia	NH_3	17.03	0.769	22.5	17.3
Methanol	CH_3OH	32.04	792	22.7	17,980
Propane	C_3H_8	44.1	2.01	50.35	101
Ethanol	C_2H_5OH	46.07	789	29.8	23,500
Gasoline	-	~114	~720	~44.5	~32,000

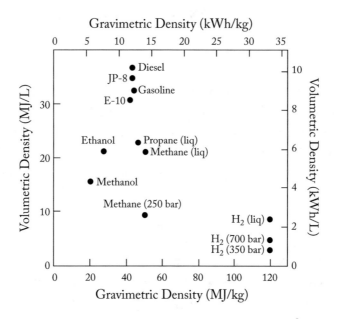

Figure 3.1: Volumetric energy density as a function of gravimetric energy density for some fuels. The energy is given by the lower heat of combustion. Image from U.S. Department of Energy; https://www.energy.gov/eere/fuelcells/hydrogen-storage.

While there has been much interest in recent years in the use of hydrogen as a storage medium for renewable energy, this element has long been, and will continue to be, an important resource in industry. Current annual world production of hydrogen is about $5{\times}10^7$ t. Table 3.3 gives a breakdown of current hydrogen use. At present, energy storage is a small fraction of the 10% listed as "other" in the table. The major use of hydrogen is for the production of ammonia, NH_3, which is used predominantly in the manufacture of fertilizers. Hydrogen also has important applications in refrigeration, plastics manufacturing, the explosives industry and water purification. The production and handling of hydrogen, as discussed below, is well established in industry.

Table 3.3: Breakdown of world hydrogen use (data adapted from https://hydrogeneurope.eu/hydrogen-uses)

Use	% World Hydrogen
Ammonia production	55
Oil refineries	25
Methanol production	10
Other	10

3.2.2   HYDROGEN PRODUCTION METHODS

Steam Reforming

About 95% of the hydrogen production worldwide is by processes that begin with fossil fuels and, ultimately, produce CO_2, as well as hydrogen. The reason for this is because the hydrogen industry is geared toward the production of industrial chemicals, and production methods involving fossil fuels are the most economical for that purpose. Most hydrogen is produced by the steam reforming of methane. This process involves a high-temperature reaction between methane and water of the form

$$CH_4 + H_2O \rightarrow CO + 3H_2. \tag{3.2}$$

The heat required for this reaction represents about 20% of the energy content of the methane, making steam reforming about 80% efficient, in terms of energy. This reaction is often followed by the reaction of carbon monoxide with water to produce more hydrogen:

$$CO + H_2O \rightarrow CO_2 + H_2. \tag{3.3}$$

This reaction, called the water-gas shift reaction, produces carbon dioxide and may have adverse environmental consequences. However, the CO_2 may be stored for use in other reactions to produce fuel as discussed next.

Electrolysis

Most of the remaining 5% of hydrogen currently produced is made by the electrolysis of water. A simple electrolysis cell is shown in Figure 3.2. The overall reaction for the production of hydrogen by electrolysis is

$$H_2O_{(l)} \rightarrow H_{2(g)} + \tfrac{1}{2}O_{2(g)}, \qquad (3.4)$$

where the subscripts (l) and (g) refer to the liquid and gas states, respectively.

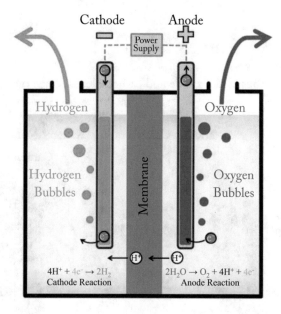

Figure 3.2: Simple electrolysis cell for the production of hydrogen gas; https://www.energy.gov/eere/fuelcells/hydrogen-production-electrolysis.

There are two designs of water electrolysis cells that are used commercially. These are the alkaline electrolyzer and the proton exchange membrane electrolyzer. The basic design of the alkaline water electrolyzer, which is the most commonly used device for water electrolysis, is shown in Figure 3.2. This electrolyzer uses an aqueous alkaline solution, typically containing potassium hydroxide (KOH) or sodium hydroxide (NaOH). A current supply is connected to two electrodes (usually nickel) separated by a polymer membrane which is permeable to water molecules and hydroxide anions. In the case of an alkaline electrolyte, the hydroxide anions travel through the membrane to the anode where they lose an electron, which then travels through the external circuit. Therefore, the reactions at the two electrodes are

$$\text{anode:} \qquad 2OH^-_{(aq)} \rightarrow H_2O_{(l)} + \tfrac{1}{2}O_{2(g)} + 2e^-, \qquad (3.5)$$

$$\text{cathode:} \qquad 2H_2O_{(l)} + 2e^- \rightarrow H_{2(g)} + 2OH^-_{(aq)}, \qquad (3.6)$$

where the subscript (aq) refers to aqueous solution. Combining Equations (3.5) and (3.6) leads to the overall reaction given in Equation (3.4).

The general design of a proton exchange membrane electrolyzer is shown in Figure 3.3. In this case, hydrogen cations travel to the cathode where they are recombined with the electron that travels through the external circuit. This leads to the reactions at the two electrodes:

$$\text{anode:} \qquad H_2O_{(l)} \rightarrow 2H^+_{(aq)} + \tfrac{1}{2}O_{2(g)} + 2e^-, \qquad (3.7)$$

$$\text{cathode:} \qquad 2H^+_{(aq)} + 2e^- \rightarrow H_{2(g)}. \qquad (3.8)$$

Again, combining the reactions at the anode and cathode leads to the overall reaction in Equation (3.4).

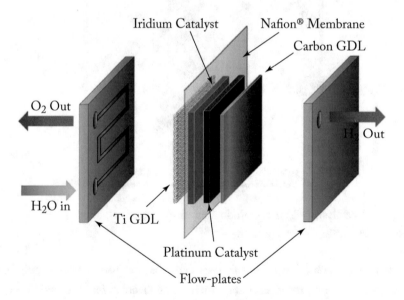

Figure 3.3: Design of a proton exchange membrane electrolyzer. Image based on Chisholm et al. (2014).

The efficiency of the electrolysis process is defined as the ratio of higher heating value of the hydrogen produced to the electrical energy input into the cell. Both alkaline electrolyzers and proton exchange membrane electrolyzers have efficiencies in the range of 70–80%.

3.2.3   HYDROGEN STORAGE METHODS

For virtually all applications, the low density of hydrogen at STP is disadvantageous for its efficient storage. In this section we look at three possible methods of making the storage of hydrogen more space efficient: high-pressure gas storage, liquid storage, and storage in solids.

High-Pressure Gas Storage

The relationship between density and pressure for hydrogen gas is shown in Figure 3.4. The ideal gas law is written as

$$PV = nRT, \tag{3.9}$$

where P is pressure, V is volume, n is the number of moles of gas, R is the universal gas constant ($R = 8.31451 \ \text{J} \cdot \text{K}^{-1} \cdot \text{mol}^{-1}$), and T is the temperature in K. Rearranging Equation (3.9) gives

$$\frac{n}{V} = \frac{P}{RT}. \tag{3.10}$$

Multiplying both sides by the molecular mass, M, gives

$$\rho = \frac{nM}{V} = \frac{MP}{RT}, \tag{3.11}$$

where ρ is the density. Thus, the ideal gas law shows a linear dependence of the density on pressure. As illustrated in Figure 3.4, at higher pressures the density does not increase as rapidly as the ideal gas law would predict. The ideal gas law is based on the assumption of noninteracting gas molecules. As the pressure increases and the density increases, the intermolecular distances decrease and the repulsive interactions between molecules causes the density to increase less rapidly as a function of pressure than the linear relationship that is predicted by the ideal gas law.

High-pressure gas storage may be appropriate for storing hydrogen for a variety of quite different applications. These include; large scale hydrogen storage, as would be appropriate for power-to-gas technology or hydrogen storage for industrial and chemical applications, hydrogen storage for stationary power systems and hydrogen storage for transportation use. For large-scale hydrogen storage, underground caverns, typically those associated with salt deposits, are used. Several large storage facilities exist in the U.S. and one sizable facility is located in the United Kingdom. These facilities are summarized in Table 3.4. For smaller-scale systems, such as hydrogen powered vehicles, pressures in the range of 35–70 MPa are typically used. Overall, the energy required to compress hydrogen to reasonable storage pressures is in the range of 10–20% of the energy content of the hydrogen, making the process 80–90% efficient.

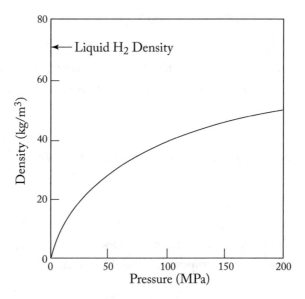

Figure 3.4: Density of hydrogen gas as a function of pressure.

Table 3.4: Underground hydrogen storage facilities in salt caverns					
Location	Volume (m3)	Mean Depth (m)	Pressure (MPa)	Hydrogen Mass (t)	Energy Stored (MJ)
Spindletop, TX, U.S.	9.06×10^5	1,340	6.8–20.2	8,230	9.86×10^8
Moss Bluff, TX, U.S.	5.66×10^6	1,200	5.5–15.2	3,690	4.42×10^8
Clemens Dome, TX, U.S.	5.80×10^5	1,000	7.0–13.7	2,400	2.92×10^8
Teeside, UK	2.10×10^5	365	4.5	810	9.72×10^7

Liquid Storage

As shown in Table 3.1, liquefying hydrogen increases its density from 0.08988–70.99 kg/m³, thereby increasing the energy density per unit volume by a factor of about 800. Obviously, in order to liquefy hydrogen gas, it is necessary to cool it to a temperature below the boiling point. However, liquefying hydrogen and maintaining it in the liquid state is not quite so simple.

In order to understand the thermodynamics of liquid hydrogen, it is important to look closely at the properties of the hydrogen molecule. The hydrogen molecule, H_2, consists of two hydrogen atoms, in this example two ^1H atoms. The proton in the nucleus of a hydrogen atom is a fermion with an intrinsic spin of 1/2. The spins of the protons in the two atoms of a hydrogen molecule can align in either of two ways, as shown in Figure 3.5. The arrangement where the two proton spins point in opposite directions is called parahydrogen and is the ground state of the molecule. The arrangement where the two proton spins point in the same direction is called orthohydrogen and is the excited state. The energy difference between the parahydrogen state and the orthohydrogen state is 0.011 eV. This means that energy is required to move a hydrogen molecule from the parahydrogen state to the orthohydrogen state and energy (in the amount of 0.011 eV) is released when a molecule in the orthohydrogen state undergoes a transition to the parahydrogen ground state. This is rather like the situation of two bar magnets configured side-by-side in the ground state with north adjacent to south and south adjacent to north. We can put the system in an excited state by rotating one of the magnets so that they are configured north to north and south to south, but this will require energy.

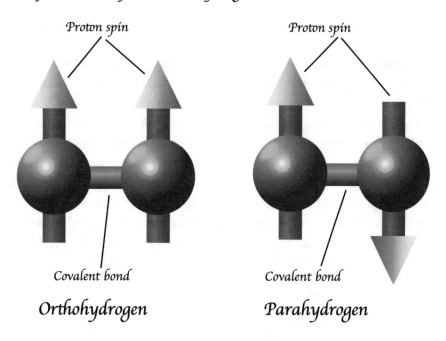

Figure 3.5: Proton spin alignment in an orthohydrogen molecule and a parahydrogen molecule; https://commons.wikimedia.org/wiki/File:Spinisomers_of_molecular_hydrogen.svg, https://creative-commons.org/licenses/by-sa/3.0/deed.en.

Room temperature (~293 K) corresponds to a thermal energy, $k_B T$, of about 0.025 eV, somewhat larger than the energy difference between the parahydrogen state and the orthohydrogen state. So, at room temperature many hydrogen molecules will be excited up into the orthohydrogen state and the equilibrium distribution will be about 25% parahydrogen and 75% orthohydrogen. As the temperature of the hydrogen is lowered, the equilibrium distribution shifts towards parahydrogen. At the boiling point of liquid hydrogen, the equilibrium distribution is about 99.8% parahydrogen and 0.2% orthohydrogen. The difficulty in producing and storing liquid hydrogen comes from the fact that if the hydrogen is cooled too quickly from room temperature to its boiling point, the system does not remain in equilibrium and the resulting liquid hydrogen can contain a significant proportion of orthohydrogen. Over time, the system tends towards equilibrium by converting orthohydrogen to parahydrogen. Each time an orthohydrogen molecule converts to a parahydrogen molecule, 0.011 eV of energy becomes available. Viewing this process from a macroscopic standpoint, the conversion of orthohydrogen to parahydrogen gives up 1.06 kJ/mol of hydrogen. This value may be compared to the latent heat of vaporization of liquid hydrogen of 0.904 kJ/mol. Thus, the conversion of molecule of orthohydrogen to parahydrogen liberates enough energy to cause the molecule to evaporate.

The liquefication of hydrogen must be done in a way that minimizes the concentration of orthohydrogen in the liquid state. This is best accomplished by cooling the hydrogen gas in the presence of a catalyst such as iron oxide or carbon, which facilitates the conversion of orthohydrogen to parahydrogen before liquefaction takes place. Overall, the liquefaction process is about 70% efficient, meaning that about 30% of the energy content of the hydrogen is needed for cooling.

Once the hydrogen has been liquefied, it is essential to prevent it from evaporating. Thus, well insulated containers are required for storing the liquid hydrogen. While the quality of insulation is important, the surface to volume ratio of the storage container is a critical factor in determining the loss rate. The larger the surface to volume ratio, the greater the loss as a fraction of the total storage capacity. This means that larger storage containers are more efficient at preserving the liquid hydrogen. Figure 3.6 shows a typical relationship between storage container volume and loss per day as a percentage of the total liquid hydrogen volume. These data show that hydrogen loss from evaporation is more problematic for small storage container, as would be appropriate for most portable applications.

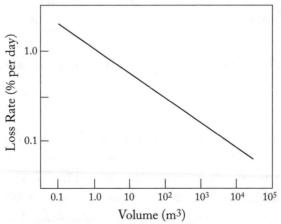

Figure 3.6: Relationship between storage container volume and liquid hydrogen loss rate per day (as a % of total volume).

Storage in Solids

Hydrogen may be stored by reacting it with other elements. The resulting material may be either a gas, a liquid or a solid. The present section considers hydrogen storage in solids. Storage in gases and liquids is discussed later in this chapter. Hydrogen storage in solids can be either by physisorption or chemisorption. We consider some commonly used and potentially promising examples of these two situations.

In the case of physisorption, the hydrogen is adsorbed onto a surface and attached by means of Van der Waals interactions. It is common to utilize materials that have a very large surface to volume ratio in order to maximize the surface area onto which the hydrogen may be adsorbed. Materials with a large amount of microstructural porosity are often most suitable. Hydrogen storage capacity can often be increased by increasing the gas pressure or by lowering the temperature.

Among the materials that have attracted interest for the physisorption of hydrogen are zeolites and nanostructured carbons. Zeolites, which are naturally occurring or synthetic aluminosilicate minerals, have an open framework structure as shown in Figure 3.7. The pores in the zeolite structure are an appropriate size for storing hydrogen and provide substantial surface area.

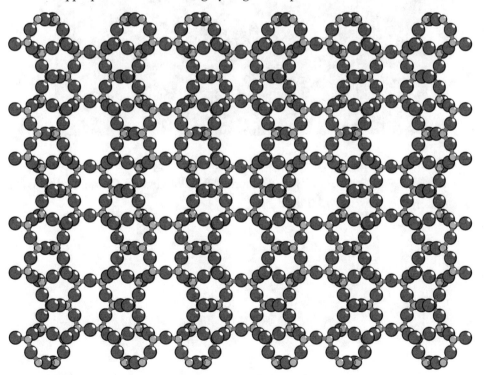

Figure 3.7: Atomic structure of the zeolite ferrierite which is a commonly used catalyst in the chemical industry. Image from Shutterstock.com, https://www.shutterstock.com/image-vector/crystal-structure-natural-zeolite-ferrierite-well-587224568.

Nanostructured carbon which is suitable for hydrogen storage by physisorption may be described in terms of the graphite structure, as shown in Figure 1.9. A single layer of the graphite structure, as shown in Figure 3.8, is the two-dimensional material known as graphene. A portion of the graphene structure may be rolled into a tube to form a carbon nanotube, as shown in Figure 3.9. If the dimensions of the tube are properly chosen, then the open volume inside the tube is appropriate for the physisorption of hydrogen.

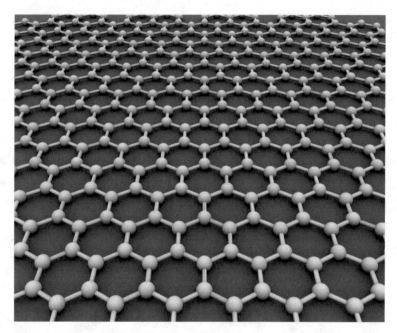

Figure 3.8: The two-dimensional hexagonal graphene structure. Image from https://en.wikipedia.org/wiki/Graphene#/media/File:Graphen.jpg, AlexanderAlUS/Wikimedia Commons, https://creativecommons.org/licenses/by-sa/3.0/deed.en.

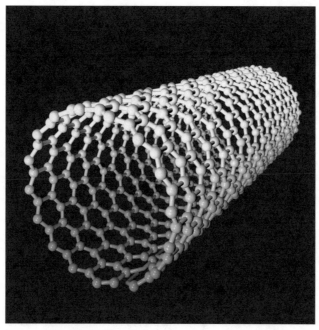

Figure 3.9: Portion of a carbon nanotube. Image from Shutterstock.com, https://www.shutterstock.com/image-illustration/carbon-nanotubes-on-dark-background-209537491.

Another approach to forming a structure with open volume from the graphene structure is to remove some of the carbon atoms, creating pentagonal cells rather than the hexagonal graphene cells. This will cause the two-dimensional graphene sheet to curve and under appropriate conditions to form a closed structure, as shown in Figure 3.10. Such molecules, as the C_{60} example shown in the figure, are referred to as fullerenes (named after the architect R. Buckminster Fuller, who developed the geodesic dome which has the same geometry as the molecule shown in the figure). Fullerenes have open volume inside the shell of carbon atoms that is suitable for storing hydrogen. Atoms inside the fullerene cage are referred to as endohedral atoms. Theoretically up to 58 endohedral hydrogen atoms can be stored inside a C_{60} fullerene molecule (Pupysheva et al., 2008). The difficulty in storing hydrogen in the fullerene molecule in Figure 3.10, is that the carbon bonds are very strong, and it is difficult to get the hydrogen atoms inside the shell, and also, once they are there, to get them out. One way of resolving this problem is to break some of the carbon bonds and attach other chemical species to form an orifice in the fullerene molecule. This is referred to as an open cage fullerene structure. Hydrogen may then pass more easily through the orifice to be stored or recovered. This is illustrated in Figure 3.11. The effectiveness of this approach may be optimized by controlling the size of the orifice and the identity of the chemical species on its rim.

Typical hydrogen storage capacities (as weight percent of the total mass) for the structures discussed above are summarized in Table 3.5.

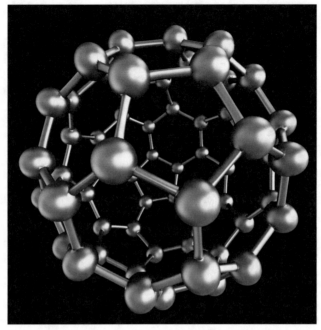

Figure 3.10: C60 buckminsterfullerene molecule. Image from Shutterstock.com, https://www.shutter-stock.com/image-illustration/metallic-fullerene-side-view-on-black-395132050?src=oG67AaPdLuEt-F2YFXW92Nw-1-29.

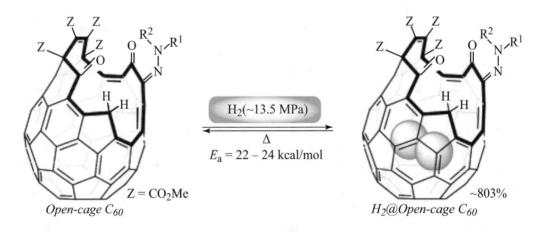

Figure 3.11: Encapsulation of a H2 molecule inside an open cage fullerene molecule. Based on Iwamatsu et al. (2005), reproduced with permission from the *Journal of Organic Chemistry*, American Chemical Society.

Table 3.5: Typical hydrogen storage capacity of some materials with open nanostructures	
Structure	**Hydrogen Capacity (mass %)**
Zeolite	2.0
Carbon nanotube	2.8
Fullerene	4.8

Hydrogen which is stored by chemisorption forms hydrides with the host material. A simple example of this storage method is the use of titanium, Ti, which, when exposed to hydrogen under appropriate conditions, forms titanium hydride, TiH_2. The hydrogen may then be extracted by decomposing the titanium hydride. This reversible reaction may be expressed as

$$Ti + H_2 \leftrightarrow TiH_2. \tag{3.12}$$

Many metals which undergo appropriate reactions with hydrogen contain one or more elements from the left side of the periodic table or one or more early transition metals. Table 3.6 summarizes some possible materials which can store hydrogen by chemisorption along with the theoretical weight percent of hydrogen in the hydride phase. Appropriate temperatures and pressures are required for effective hydrogen absorption and desorption.

Table 3.6: Some metal hydrides that are of interest as possible hydrogen storage materials	
Metal Hydride	**Weight Percent Hydrogen**
LaNi5H6	1.4
TiFeH2	1.9
TiH2	4.0
NaAlH4	7.4
MgH2	7.6
LiAlH4	10.5
LiH	12.5

3.2.4   FUEL CELL TECHNOLOGY

As mentioned above, one method of recovering energy that is stored in hydrogen is by combustion, as described in Equation (3.1). Alternately, energy may be obtained from hydrogen by combining hydrogen with oxygen in a fuel cell. There are a variety of designs of fuel cells. These differ in terms of the electrolyte that is used and the ion which is transferred between the anode and cathode. They also require different operating conditions, such as temperature. Table 3.7 summarizes some of the relevant features of the major types of fuel cells. The alkaline fuel cell and the proton exchange membrane fuel cell are, perhaps, the most commonly used and are analogous in their operation to the alkaline and proton exchange membrane electrolyzer, respectively, as described above. The alkaline fuel cell has been used by NASA as a source of energy in spacecraft, while the proton electrolyte membrane fuel cell is the most suitable for motor vehicle applications. The operation of these two types of fuel cells is illustrated in Figures 3.12 and 3.13, respectively. An inspection of the figures shows that the reactions at the anode and cathode in the alkaline fuel cell are the reverse of the reactions shown in Equations (3.5) and (3.6) while the reactions at the anode and cathode in the proton exchange membrane fuel cell are the reverse of the reactions shown in Equations (3.7) and (3.8). Both sets of reactions lead to an overall reaction where hydrogen and oxygen are combined:

$$H_2 + \tfrac{1}{2}O_2 \rightarrow H_2O. \tag{3.13}$$

Table 3.7: Typical characteristics of some common fuel cells (data adapted from https://www.energy.gov/sites/prod/files/2015/11/f27/fcto_fuel_cells_fact_sheet.pdf)

Type	Typical Electrolyte	Ion Transferred	Power Output Range (kW)	Operating Temperature (°C)	Cell Efficiency (%)
Alkaline	KOH	OH^-	1 kW–100 kW	< 100	60
Proton exchange membrane	Polymer membrane	H^+	1 kW–100 kW	< 120	60
Phosphoric acid	Molten H_3PO_4	H^+	5 kW–400 kW	150–200	40
Molten carbonate	Molten K_2CO_3	CO_3^{2-}	300 kW–3 MW	600–700	50
Solid oxide	ZrO_2	O^{2-}	1 kW–2 MW	500–1,000	60

Some fuel cells may also be designed to operate using light hydrocarbons rather than pure hydrogen as a fuel. For example, solid oxide fuel cells can use methane as a fuel according to the reaction

$$CH_4 + 2O_2 \rightarrow CO_2 + 2H_2O. \tag{3.14}$$

In addition, proton exchange membrane fuel cells can be designed to operate using methanol as a fuel according to the reaction

$$2CH_3OH + 3O_2 \rightarrow 2CO_2 + 4H_2O, \tag{3.15}$$

although typically at a reduced efficiency of about 35%. In both cases, as shown in Equations (3.14) and (3.15), CO_2 is a by-product of the reaction. If the methane or methanol is produced using carbon-free methods as described below, then the overall process will be carbon-neutral.

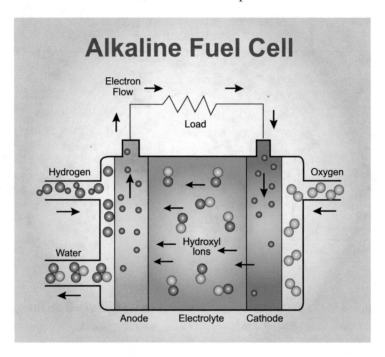

Figure 3.12: Operation of the alkaline fuel cell. Image from Shutterstock.com, https://www.shutterstock.com/image-vector/alkaline-fuel-cells-consume-hydrogen-pure-408141571?src=4iQ8DWXVVXXi-xKpJFIOdA-1-18.

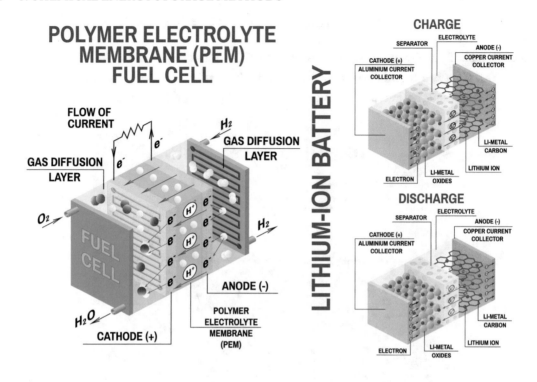

Figure 3.13: Operation of the proton exchange membrane fuel cell. Image from Shutterstock.com, https://www.shutterstock.com/search?search_source=base_landing_page&language=en&searchterm=-fuel+cell&image_type=all.

3.2.5   APPLICATIONS OF HYDROGEN ENERGY STORAGE

Power-to-Gas

The term "power-to-gas" refers to the process of using excess electricity (generated at times of low demand) to produce hydrogen by electrolysis. The hydrogen is then injected into the natural gas grid to supplement the energy content of the natural gas. The concentration of the hydrogen in natural gas is limited to 2%. While it is common to use this approach for excess electricity produced by carbon-free renewable technologies, such as wind or solar photovoltaics, the subsequent use of natural gas (from fossil fuel sources) contributes to greenhouse gas emissions. The efficiency of power-to-gas technology depends on the efficiency of the electrolysis and the efficiency of the compression of the gas. Hydrogen may be injected into the natural gas network with minimal compression, it may be injected into a natural gas pipeline at a pressure of about 8 MPa or it may be compressed for storage. Typical efficiencies for these scenarios are given in Table 3.8.

Table 3.8: Power-to-gas efficiencies for hydrogen using different compression conditions, (data adapted from https://www.greenpeace-energy.de/fileadmin/docs/sonstiges/Greenpeace_Energy_Gutachten_Windgas_Fraunhofer_Sterner.pdf)

Use	Pressure	Net Efficiency
Direct injection into natural gas grid	350 kPa	64–77%
Injection into natural gas pipeline	8 MPa	57–73%
Storage	20 MPa	54–72%

Hydrogen produced by electrolysis from excess electricity may also be converted into methane as discussed in Section 3.3 for injection into the natural gas grid or used to upgrade the quality of biogas.

Stationary Power Systems

Hydrogen fuel cell systems can serve the same purpose as battery backup systems (e.g., Figure 1.31) to provide power in the event of a grid outage for facilities such as data communication centers, police stations, hospitals, etc. Commercially available systems, like battery backup grid storage systems (see Figure 1.34), are often constructed in shipping containers, as shown in Figure 3.14.

Figure 3.14: Hydrogen fuel cell power backup system. Image reproduced with permission from https://www.energy.gov/eere/success-stories/articles/eere-success-story-fuel-cell-generators-prove-they-can-save-energy-and.

In remote locations which are not connected to the power grid, hydrogen may be used as an energy storage means to make effective use of renewable energy sources and to meet demands during peak periods. In 2010, a pilot program which took this approach was implemented in the village of Ramea (population ~450) which is located on Northwest Island about 8 km off the south coast of Newfoundland and Labrador in Canada. A diagram of the design of the Ramea power system is shown in Figure 3.15. Wind turbines supplement base load power provided by a diesel generator. Excess electricity is used to produce hydrogen by electrolysis, which is then compressed for storage. Power to meet demand when direct output from wind turbines and/or diesel generation is not sufficient is provided by electricity generated by an internal combustion hydrogen generator using stored hydrogen.

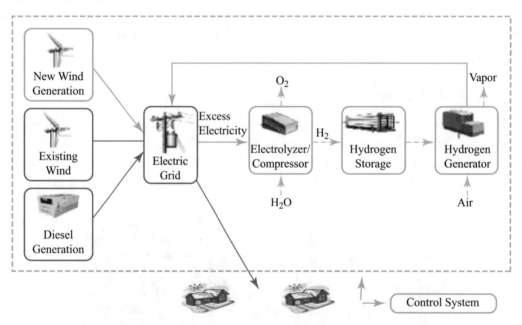

Figure 3.15: Layout of electric grid for Ramea, Newfoundland and Labrador. Image based on https://www.nrcan.gc.ca/energy/renewable-electricity/wind/7319.

Transportation

Perhaps the most extensive research and development into the use of hydrogen as an energy storage method has been in the area of transportation. Two approaches have been taken for the direct use of hydrogen as a transportation fuel: burning hydrogen in an internal combustion engine (sometimes called a H2ICE) and producing electricity from hydrogen via a fuel cell to drive an electric motor.

Here we review some of the more notable hydrogen powered vehicles that have been produced in recent years.

While a conventional gasoline powered internal combustion engine requires only fairly simple modifications to run on hydrogen, the implementation of a suitable hydrogen fuel storage system is somewhat more involved. Beginning in 2003, Mazda produced a limited-production RX-8 with a rotary (Wankel) engine that ran on hydrogen (the RX-8 Hydrogen RE). This vehicle used hydrogen gas stored at a pressure of 35 MPa and is shown in Figure 3.16. At a pressure of 35 MPa, Figure 3.4 shows that hydrogen gas has a density of slightly over 20 kg/m^3. The hydrogen storage tank had a volume of 110 L, yielding a total of 2.4 kg of stored hydrogen fuel. This mass of hydrogen gave the RX-8 Hydrogen RE a range of about 100 km (consistent with an energy requirement from the fuel of about 3 MJ/kg, as shown in Table 1.5 and the lower heat of combustion of hydrogen, 120 MJ/kg from Table 3.1). The Mazda RX-8 Hydrogen RE was a hybrid vehicle (as are virtually all hydrogen powered vehicles) which used gasoline as well as hydrogen as a fuel, since the Wankel engine can burn either hydrogen or gasoline. Table 3.9 summarizes the specifications of the RX-8 Hydrogen RE for the hydrogen and gasoline modes. It is clear, as a result of the greater specific energy density of gasoline compared to hydrogen, that the vehicle range is greater for the gasoline mode of operation than for the hydrogen mode. More recently (beginning in 2005), Mazda produced a limited-production tri-fuel hybrid vehicle, the Premacy Hydrogen RE, as shown in Figure 3.16, with a Wankel engine that burns both hydrogen and gasoline, as well as an electric drive motor operated by nickel metal hydride batteries.

Figure 3.16: **Mazda RE-8 Hydrogen RE.** Image from Taisyo at https://en.wikipedia.org/wiki/ Mazda_RX-8_Hydrogen_RE#/media/File:Mazda_RX8_hydrogen_rotary_car_1.jpg.

Table 3.9: Specifications of the Mazda RX-8 Hydrogen RE	
Specification	Value
Hydrogen fuel volume and pressure	110 L at 35 MPa
Engine output (hydrogen)	80 kW
Vehicle range (hydrogen)	100 km
Gasoline volume	61 L
Engine output (gasoline)	154 kW
Vehicle range (gasoline)	530 km

Figure 3.17: **Mazda Premacy Hydrogen RE Hybrid;** https://commons.wikimedia.org/wiki/
File:Mazda_Premacy_HRE_Hybrid.JPG.

The BMW Hydrogen 7 (Figure 3.18) was produced from 2005–2007 and used a 6.0 L V-12 internal combustion engine (Figure 3.19) modified to burn both hydrogen and gasoline. The BMW Hydrogen 7 is probably unique among hydrogen powered passenger vehicles, as it used liquid hydrogen, rather than high pressure hydrogen gas, as a fuel. The liquid hydrogen was stored in a vacuum insulated tank. Specifications for the BMW Hydrogen 7 are given in Table 3.10.

Figure 3.18: The BMW Hydrogen 7. Image from Sachi Gahan at https://commons.wikimedia.org/wiki/File:BMW_Hydrogen_7_at_TED_2007.jpg.

Figure 3.19: 6.0-L V-12 internal combustion engine used in the BMW Hydrogen 7. Image from Claus Ableiter at https://commons.wikimedia.org/wiki/File:Motor_Hydrogen_7.JPG.

Table 3.10: Specifications of the BMW Hydrogen 7	
Specification	Value
Hydrogen fuel volume (liquid)	~110 L
Engine output (hydrogen)	191 kW
Vehicle range (hydrogen)	201 km
Gasoline volume	73.8 L
Engine output (gasoline)	191 kW
Vehicle range (gasoline)	480 km

Many automobile manufacturers have worked on the development of fuel cell vehicles in recent years. At present (2019), four vehicles are in production and are available (in limited quantities and in limited areas) for the general public to purchase and/or lease. These vehicles are the Hyundai Tuscon FCEV, the Toyota Mirai (Figure 3.20), the Honda Clarity (Figure 3.21), and the Hyundai Nexo (Figure 3.22). Purchase price for these vehicles, where they are sold, is typically in the $60,000–$70,000 USD range and, at present, they are sold at a loss to the manufacturer. All vehicles use hydrogen stored as high-pressure gas at 70 MPa and are fuel cell/battery hybrids. As the fuel cell drive system includes electric motors to drive the wheels, it is straightforward to add batteries for additional energy storage from regenerative braking. Table 3.11 provides specifications for these four vehicles.

Figure 3.20: **Toyota Mirai fuel cell vehicl;** https://commons.wikimedia.org/wiki/File:Toyota_FCV_Concept.jpg.

Figure 3.21: **Honda Clarity fuel cell vehicle.** Image from Alexander Migi at https://commons.wikimedia.org/wiki/File:Honda_Clarity_Fuel_Cell_IMG_0301.jpg.

Figure 3.22: **Hyundai Nexo fuel cell vehicle;** https://commons.wikimedia.org/wiki/File:Hyundai_Nexo_-_f_16032019.jpg, https://creativecommons.org/licenses/by-sa/3.0/de/deed.en.

Table 3.11: Specifications of currently available fuel cell vehicles (as of January 2019)						
Make	Model	First Year Produced	Hydrogen Mass (kg)	Power (kW)	Range (km)	Battery Type
Hyundai	Tuscon FCEV	2014	5.64	100	594	Li-ion
Toyota	Mirai	2015	5.0	113	502	NiMH
Honda	Clarity	2016	5.4	103	589	Li-ion
Hyundai	Nexo	2018	6.3	120	600	Li-ion

As a result of automobile manufacturer's initiatives to develop hydrogen powered vehicles, hydrogen filling stations have been established; see Figure 3.23. The majority of these stations are in California, Europe and Japan and mostly provide compressed hydrogen gas. Very few stations provide liquid hydrogen.

Figure 3.23: Hydrogen fueling station in Washington, DC. Image from Rob Crandall, Shutterstock. com, https://www.shutterstock.com/image-photo/washington-dc-usa-march-29-2006-558258223?s-rc=y2YRYXj9iiqNySE3Khti_A-1-4.

3.2.6 EFFICIENCY OF HYDROGEN ENERGY STORAGE

A flow diagram for possible hydrogen use as an energy storage medium is shown in Figure 3.24. This system may be compared to, for example, a similar system using batteries for energy storage, as illustrated in Figure 1.29. Since energy conversions are always less than 100% efficient, the overall efficiency of the system should be considered. The efficiency of batteries as a means of energy storage, as discussed in Chapter 1, is very high and the net charge/discharge efficiency for grid storage can be around 90%.

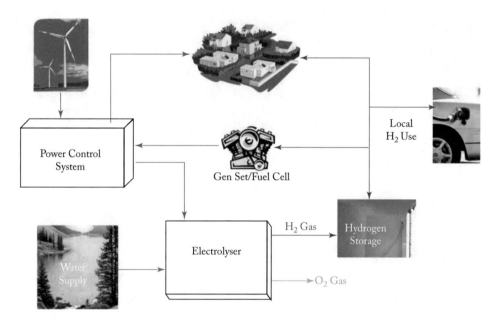

Figure 3.24: Diagram of hydrogen energy storage system for energy provided by renewable sources, Based on https://commons.wikimedia.org/wiki/File:Wind_hydrogen.JPG.

Based on the discussion above, the efficiency of hydrogen energy storage as outlined in Figure 3.24 may be considered. We may express the net efficiency of storage and energy recovery as

$$\eta_{net} = \eta_{production} \times \eta_{storage} \times \eta_{recovery},\qquad(3.16)$$

where

- $\eta_{production}$ is the efficiency of steam reforming or electrolysis,

- $\eta_{storage}$ is the efficiency of compression or liquefaction, and

- $\eta_{recovery}$ is the efficiency of recovering energy that is stored in hydrogen.

Here the most common methods of hydrogen production and storage are considered. On the basis of the above discussion, we use typical efficiencies of 70–80% for production and 70–90% for storage giving a net production/storage efficiency of 49–72%. The efficiency of energy recovery is much more variable, depending on the intended use of the energy and the technology utilized. Based on Figure 3.24, we consider the examples shown in Table 3.12. It is clear that efficiency is a concern for hydrogen energy storage technology.

Table 3.12: Estimated efficiency of energy systems involving hydrogen storage. The efficiency of alkaline and proton exchange membrane fuel cells (60%) is used. The efficiency of an internal combustion engine was assumed to be 20%

Application	Energy Recovery Method	$\eta_{recovery}$ (%)	η_{net} (%)
Electricity	Fuel cell	60	29–43
Electricity	H2ICE/generator	20	10–14
Vehicle	Fuel cell	60	29–43
Vehicle	H2ICE	20	10–14
Space heating	Combustion	90	44–65

3.3   METHANE

Methane, CH_4, is the simplest hydrocarbon and is the major component of natural gas. It is more convenient than hydrogen for many applications because the storage, transport, and combustion infrastructure is already in place in the form of storage tanks, pipelines, trucking, and natural gas-fired power plants. Methane may be produced from hydrogen by the Sabatier reaction:

$$4H_2 + CO_2 \rightarrow CH_4 + 2H_2O. \tag{3.17}$$

This reaction takes place in the presence of a catalyst such as nickel, ruthenium, or alumina, and requires temperatures in the range of 300–400°C. About 8% of the energy content of the hydrogen is lost in the process. Efficiencies for methane production from hydrogen prepared by electrolysis can, therefore, be related to the electrolysis efficiencies given in Table 3.8. Synthetic natural gas, produced in this way, may be substituted, without limit, into the natural gas grid as described above for hydrogen. Carbon dioxide that is sequestered from fossil fuel combustion or that is produced by steam reforming may be used for the production of methane. If the hydrogen is produced in a carbon-free manner (i.e., electrolysis using renewable energy) then the use of the resulting methane is carbon-neutral. Methane may also be used directly as a fuel in some types of fuel cells, specifically solid oxide fuel cells.

3.4   METHANOL

Hydrogen may be combined with carbon dioxide to produce methanol according to the reaction

$$3H_2 + CO_2 \rightarrow CH_3OH + H_2O. \qquad (3.18)$$

Methanol may also be formed from the reaction of hydrogen with carbon monoxide (which is produced during steam reforming of methane to produce hydrogen) by the reaction

$$2H_2 + CO \rightarrow CH_3OH, \qquad (3.19)$$

or by the oxidation of methane

$$2CH_4 + O_2 \rightarrow 2CH_3OH. \qquad (3.20)$$

These reactions take place at elevated temperature and are generally facilitated by the presence of an appropriate catalyst. Overall methanol production efficiency can be as high as 75% (Bozzano and Manenti 2016).

Methanol is a convenient, direct substitute for gasoline in internal combustion engines and burns according to the reaction in Equation (3.15) where the carbon dioxide by-product balances the CO_2 used to produce the methanol according to, for example, Equation (3.18). Thus, as for the production of methane from hydrogen, this process is carbon-free if the hydrogen is produced without carbon emissions.

Minimal modifications are required to convert an internal combustion engine from gasoline to methanol operation. As a liquid, methanol eliminates the storage difficulty that accompanies gaseous fuels such as hydrogen and methane. As seen in Table 3.1, methanol's energy density per unit volume is about half that of gasoline, meaning a greater volume of fuel is needed to provide the same vehicle range. However, methanol provides greater engine power output than gasoline and thermodynamic efficiencies as high as 42% have been observed in properly optimized methanol internal combustion engines (Vancoillie, 2013). Methanol has some clear advantages over gasoline; it is less flammable, as shown in Table 3.13, and is more biodegradable. It is, therefore, safer and presents less environmental concerns. Other factors that need to be considered when using methanol as a fuel, are its hygroscopicity, requiring that fuel containers be sealed to eliminate moisture, and its tendency to contain halide impurities, which can corrode engine components.

Table 3.13: Flammability of some common fuels. The flammability range is the range of concentrations of the fuel in air that is flammable. The autoignition temperature is the temperature at which the fuel with spontaneously ignite without a source of ignition

Fuel	Flammability Range (%)	Autoignition Temperature (°C)
Gasoline	1–6	222
Methane	5–15	534
Hydrogen	4–75	585
Methanol	7–36	585
Ammonia	15–28	651

The George Olah Renewable Methanol Plant in Svartsengi, Iceland is operated by Carbon Recycling International and produces methanol by the reaction shown in Equation (3.18) using hydrogen produced by electrolysis with electricity provided by carbon-free sources (hydroelectric and geothermal energy). This facility is the world's largest renewable methanol plant and currently produces about 5 million liters of methanol annually. It was named in honor of George Olah, 1994 Chemistry Noble Prize winner, who has promoted methanol as a sustainable energy carrier for the future. Chinese automobile manufacturer Geely (which owns Volvo) has recently begun testing a fleet of methanol powered vehicles in Iceland where the fuel is marketed under the name "Volcanol" (see Figure 3.25).

Figure 3.25: Geely Emgrand EC7 methanol vehicle with the George Olah Renewable Methanol Plant in Svartsengi, Iceland in the background. Image used with permission from Geely Auto; https://igpmethanol. com/2017/08/05/volvos-chinese-owner-tests-methanol-cars-in-iceland-for-european-markets/ZZZ.

3.5   AMMONIA

3.5.1   PROPERTIES OF AMMONIA

As noted in Table 3.3, the majority of world hydrogen production is used for the purpose of making ammonia. Ammonia can be produced through the reaction of hydrogen with atmospheric nitrogen:

$$3H_2 + N_2 \rightarrow 2NH_3. \tag{3.21}$$

This reaction, known as the Haber-Bosch process, occurs at about 450°C and a pressure of 10 MPa, over a magnetite catalyst. Equation (3.21) shows that ammonia may be thought of as a means of storing hydrogen by chemisorption.

Ammonia has a number of industrial applications including the manufacture of fertilizers and use as a refrigerant. However, ammonia has two possible uses as a fuel; first, as a combustible fuel to replace petroleum products and, second, as a medium for hydrogen storage. We begin with some basic properties of ammonia.

Table 3.14 lists some properties of ammonia. It is seen that ammonia at standard temperature and pressure is a gas. The energy density per unit volume is, therefore, very low (as is the case for gaseous hydrogen). The boiling point of ammonia at 1 atmosphere is 240 K or -33°C (compared to 20.3 K for hydrogen). It is, therefore, much easier to liquefy ammonia than hydrogen, and this process requires much less energy. Figure 3.26 shows the temperature dependence of the vapor pressure of ammonia. It is seen that the vapor pressure at 20°C is about 6,500 mm Hg or about 8.5 atmospheres (actual value is 857 kPa). This is a much more manageable pressure than that which is used for high pressure hydrogen storage (35–70 MPa). Thus, ammonia may be transported as a liquid at temperatures that are readily obtained with conventional refrigeration or in pressurized tanks at relatively modest pressures.

Table 3.14: Some properties of ammonia. Note: Energy is the higher heat of combustion		
Property	**Value**	**Units**
Boiling point (1 atm)	240	K
Liquid density	771	kg/m^3
Gas density	0.769	kg/m^3
Energy	22.5	MJ/kg
Energy density (gas)	17.3	MJ/m^3
Energy density (liquid)	17,348	MJ/m^3

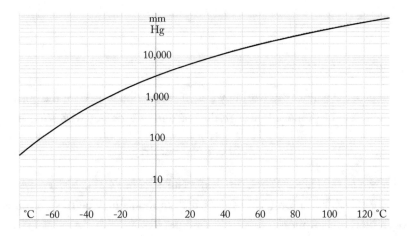

Figure 3.26: Vapor pressure of ammonia as a function of temperature; https://commons.wikimedia.org/wiki/File:LogAmmoniaVaporPressure.png.

In terms of energy density, it is appropriate to think of ammonia as a liquid fuel, and its energy density is shown in comparison to some other liquid fuels in Table 3.15. While the energy density of ammonia per unit volume is only about half that of gasoline, it is similar to that of methanol. The combustion process for ammonia is

$$4NH_3 + 5O_2 \rightarrow 4NO + 6H_2O. \tag{3.22}$$

It is important to note that pure hydrogen and ammonia are the only fuels that do not contain carbon and, therefore, do not produce any CO_2 during combustion.

Table 3.15: Higher heat of combustion per volume for some liquid fuels	
Fuel	Energy (MJ/L)
Gasoline	33.0
Liquid hydrogen	10.1
Methanol	18.0
Ammonia	17.3

3.5.2   AMMONIA INTERNAL COMBUSTION ENGINES

The information above suggests that ammonia may be an attractive alternative to the use of fossil fuels. Ammonia may be used directly as a fuel in an internal combustion engine and because of its properties can be ignited either by spark ignition, as is used for a gasoline engine, or by compression ignition, as is used for a diesel engine. Because of its low flammability, as indicated in Table 3.13, ammonia is, in some respects, safer than gasoline or hydrogen. However, there are some serious drawbacks and concerns about the use of ammonia as a fuel in an internal combustion engine.

Because of its low flammability, ammonia is not as easy to ignite as fuels like gasoline. This leads to difficulties in maintaining combustion under all engine speeds and loads. There are several approaches to dealing with this difficulty. The simplest approach is to add a combustion promoter to the ammonia fuel. This combustion promoter can be a fossil fuel, such as gasoline, or, in order to eliminate net atmospheric carbon emissions, it can be biodiesel. This approach may be used in conjunction with either spark or compression ignition, but typically requires two fuel distribution systems in order to match the fuel mixture to the engine operating conditions. Hydrogen may also be used as a combustion promoter. The hydrogen can be obtained by cracking a small amount of the ammonia from the on-board fuel supply by the reaction

$$2NH_3 \rightarrow N_2 + 3H_2. \tag{3.23}$$

Cracking is discussed below for fuel cell vehicles. This approach is best utilized for engines with spark ignition and eliminates the release of greenhouse gases. However, the addition of a cracking system adds to the weight and complexity of the engine and it is difficult to adjust the fuel mixture to effectively match all operating conditions.

Other difficulties that are related to the use of ammonia include its toxicity. Concentrations of ammonia in air of a few hundred ppm can be a severe irritant and concentrations of a few thousand ppm can be fatal. It is, therefore, essential that vehicle fueling be done in a well-controlled way and that fuel tanks are highly resistant to leakage.

Ammonia is highly corrosive, and the fuel system and engine components must be designed to deal with this problem. Ammonia is also hygroscopic and exposure to atmospheric moisture must be avoided to eliminate the possibility of contamination.

Although the combustion of ammonia does not produce carbon dioxide or other greenhouse gases, it does produce NO as shown in Equation (3.22). NO reacts with oxygen in the atmosphere to produce NO_2, which is a major component of smog in urban areas.

3.5.3   HYDROGEN STORAGE IN AMMONIA

Ammonia may also be viewed as a means of hydrogen storage. The hydrogen is chemisorbed by nitrogen to form ammonia. The weight percent of hydrogen in ammonia, 17.8 wt%, is greater than for other hydrogen storage media; see Table 3.6. For vehicle use, ammonia may be conveniently carried on-board and cracked to yield a source of hydrogen, emitting only nitrogen, as shown in Equation (3.23). The cracking reaction is the reverse of the process shown in Equation (3.21). This reaction is endothermic, and energy must be supplied in order for it to proceed. The energy is supplied in the form of heat and the decomposition of ammonia is done at temperatures typically above about 600°C in the presence of a metallic catalyst. A schematic of a cracking system is shown in Figure 3.27. The hydrogen, thus produced, may be used as a fuel for a proton exchange membrane fuel cell to produce electricity for vehicle propulsion.

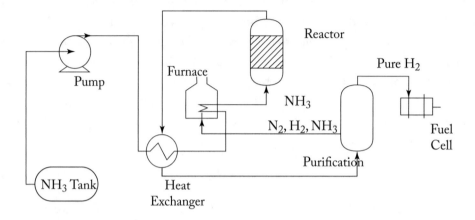

Figure 3.27: Simplified schematic of an ammonia cracking system. Based on Thomas and Parks (2006), used with permission from energy.gov; https://www.energy.gov/sites/prod/files/2015/01/f19/fcto_nh3_h2_storage_white_paper_2006.pdf.

There are several important considerations concerning the viability of this approach. First, proton exchange membrane fuel cells cannot tolerate ammonia impurities in the hydrogen in excess of about 0.1 ppm. Figure 3.28 shows the results of a model for ammonia concentration in hydrogen produced by cracking as a function of cracking temperature and ammonia pressure. It is clear that, even at 1 atmosphere and a cracking temperature of 900°C, the ammonia impurity concentration is well above the acceptable limit. Thus, as shown in Figure 3.27, a suitable purification system to significantly reduce NH3 contamination in the hydrogen is necessary.

From a practical standpoint, the cracking apparatus is heavy and bulky and start-up time for the cracker to reach operating temperature may be an adverse factor for vehicular use. Therefore, the use of ammonia as a hydrogen storage medium may, at present, be best suited for stationary facilities.

The use of ammonia, either as a direct fuel for an internal combustion engine or as a hydrogen storage medium, must be considered in the context of overall system efficiency. While, at present, a detailed efficiency analysis is lacking, it is clear from the number of energy conversions involved, as well as a comparison with the direct use of hydrogen in internal combustion engines or in fuel cells, as summarized in Table 3.13, that the efficiency of using ammonia requires serious consideration.

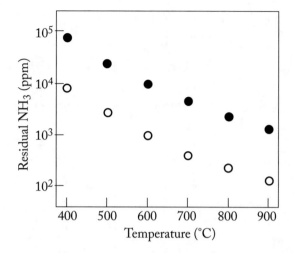

Figure 3.28: Residual NH3 after cracking as a function of temperature for ammonia pressures of 10^5 Pa (open circles) and 10^6 Pa (closed circles). Data are from Thomas and Parks (2006).

Bibliography

Alotto, P., Guarnieri, M., and Moro, F. (2014). Redox flow batteries for the storage of renewable energy: A review. *Renewable and Sustainable Energy Reviews* 29, 325–335. DOI: 10.1016/j.rser.2013.08.001. 17

Ashcroft, N. W. and Mermin, N. D. (1976). *Solid State Physics*. Boston, MA: Cengage.

Bozzano, G. and Manenti, F. (2016). Efficient methanol synthesis: Perspectives, technologies and optimization strategies. *Progress in Energy and Combustion Science* 56, 71–105. DOI: 10.1016/j.pecs.2016.06.001.

Burns, G. (1991). *High-Temperature Superconductivity: An Introduction*, 1st ed. New York: Academic Press.

Ciceron, J., Badel, A., and Tixador, P. (2017). Superconducting magnetic energy storage and superconducting self-supplied electromagnetic launcher, *The European Physical Journal Applied Physics* 80, 20901. DOI: 10.1051/epjap/2017160452.

Chisholm, G., Kitson, P. J., Kirkaldy, N. D., Bloor, L. G., and Cronin, L. (2014). 3D printed flow plates for the electrolysis of water: an economic and adaptable approach to device manufacture. *Energy and Environmental Science* 7, 3026-3032 DOI: 10.1039/C4EE01426J. 64

Chisholm, G. and Cronin, L. (2016). Hydrogen from water electrolysis. In Letcher, T. M., ed., *Storing Energy—with Special Reference to Renewable Energy Sources*. Amsterdam: Elsevier. Chapter 16 pp. 315–343. DOI: 10.1016/B978-0-12-803440-8.00016-6.

Connor, A. (2017). Batteries for energy storage. In Crawley, G. M., ed., *Energy Storage*. Singapore: World Scientific. Chapter 4, pp. 117–165. DOI: 10.1142/9789813208964_0004.

Crotogino, F. (2016). Larger scale hydrogen storage. In Letcher, T. M., ed., *Storing Energy—with Special Reference to Renewable Energy Sources*. Amsterdam: Elsevier. Chapter 20, pp. 411–429. DOI: 10.1016/B978-0-12-803440-8.00020-8.

da Rosa, A. V. (2013). *Fundamentals of Renewable Energy Processes*, 3rd ed. Oxford: Academic Press. DOI: 10.1016/B978-0-12-397219-4.00017-5.

Doetsch, C. and Burfeind, J. (2016). Vanadium redox flow batteries. In Letcher, T. M., ed., *Storing Energy—with Special Reference to Renewable Energy Sources*. Amsterdam: Elsevier, Chapter 12, pp. 227–246. DOI: 10.1016/B978-0-12-803440-8.00012-9.

Dunlap, R. A. (2018). *Novel Microstructures for Solids, IOP Concise Physics Series*. San Rafael, CA: Morgan & Claypool. DOI: 10.1088/2053-2571/aae653.

Dunlap, R. A. (2019). *Sustainable Energy*, 2nd ed. Boston, MA: Cengage.

Gallagher, K. G., Dees, D. W., Jansen, A. N., Abraham, D. P., and Kang, S. H. (2012). A volume averaged approach to the numerical modeling of phase-transition intercalation electrodes presented for LixC6. *Journal of the Electrochemical Society* 159, A2029–A2037. DOI: 10.1149/2.015301jes. 10

Gallop, J. C. (1990). *SQUIDS, The Josephson Effects and Superconducting Electronics*. Boca Raton, FL: CRC Press.

Ghoniem, A. F. (2011). Needs, resources and climate change: Clean and efficient conversion technologies. *Progress in Energy and Combustion Science* 37, 15–51. DOI: 10.1016/j.pecs.2010.02.006. 43

Gu, W. T., Wei, L., and Yushin, G. (2017). Capacitive energy storage. In Crawley, G. M., ed., *Energy Storage*. Singapore: World Scientific, pp. 167-214. DOI: 10.1142/9789813208964_0005.

Holla, R. V. (2015). Energy Storage methods–Superconducting magnetic energy storage–A review. *The Journal of Undergraduate Research* 5, 49–54. DOI: 10.5210/jur.v8i1.7540.

Huggins, R. (2016). *Energy Storage: Fundamentals, Materials and Applications*, 2nd ed. Berlin: Springer.

Hwang, J. Y., Myung, S. T., and Sun, Y. K. (2017). Sodium-ion batteries: present and future. *Chemical Society Reviews* 46, 3529–3614. DOI: 10.1039/C6CS00776G. 15

Huggins, R. (2016). *Energy Storage: Fundamentals, Materials and Applications*, 2nd ed. Berlin: Springer.

Iwamatsu, S., Murata, S., Andoh, Y., Minoura, M., Kobayashi, K., Mizorogi, N., and Nagase, S. (2005). Open-cage fullerene derivatives suitable for the encapsulation of a hydrogen molecule. *Journal of Organic Chemistry* 70, 4820–4825. DOI: 10.1021/jo050251w. 72

Johnson, V. H., Pesaran, A. A., and Sack, T. (2001). Temperature-dependent battery models for high-power lithium-Lon batteries. *Report NREL/CP-540-28716* from National Renewable Energy Laboratory. Available at https://www.nrel.gov/docs/fy01osti/28716.pdf. 26

Kleiner, R. and Buckel, W. (Huebener, R., translator) (2016). *Superconductivity: An Introduction*, 3rd ed. Hoboken, NJ: Wiley. DOI: 10.1002/9783527686513.

Kruger, P. (2006). *Alternative Energy Resources: The Quest for Sustainable Energy*. Hoboken, NJ: Wiley.

Linden, D. and Reddy, T. B, eds. (2002). *Handbook of Batteries*, 3rd ed. New York: McGraw-Hill.

Masters, G. M. (2013). *Renewable and Efficient Electric Power Systems*, 2nd ed. Hoboken, NJ: Wiley.

Nagaya, S., Hirano, N., Moriguchi, H., and Shikimachi, K. (2006). Field test results of the 5 MVA SMES System for bridging instantaneous voltage dips. *IEEE Transactions on Applied Superconductivity* 16(2). DOI: 10.1109/TASC.2005.864359. 56

Ngô, C. and Natowitz, J. B. (2009). *Our Energy Future: Resources, Alternatives and the Environment.* Hoboken, NJ: Wiley.

Olah, G. A., Goeppert, A., and Surya Prakash, S. K. (2009). *Beyond Oil and Gas: The Methanol Economy*, 2nd ed. Weinheim, Germany: Wiley-VHC. DOI: 10.1002/9783527627806.

Patel, M. R. (2006). *Wind and Solar Power Systems: Design, Analysis and Operation*, 2nd ed. Boca Raton, FL: Taylor & Francis. DOI: 10.1201/9781420039924.

Pupysheva, O. V., Farajian, A. A., and Yakobson, B. I. (2008). Fullerene nanocage capacity for hydrogen storage. *Nano Letters* 8, 767–774. DOI: 10.1021/nl071436g. 71

Ray, P. J. (2015). Structural investigation of La2-xSrxCuO4+y - Following staging as a function of temperature. Master's thesis, Niels Bohr Institute, Faculty of Science, University of Copenhagen. Copenhagen, Denmark. 52

Romm, J. J. (2005). *The Hype About Hydrogen: Fact and Fiction in the Race to Save the Climate.* Washington, DC: Island Press.

Santos, D. M. F., Sequeira, C. A. C., and Figueiedo, J. L. (2013). Hydrogen production by alkaline water electrolysis. *Química Nova* 36, 1176–1193. DOI: 10.1590/S0100-40422013000800017.

Scott, D. S. (2008). *Smelling Land: The Hydrogen Defense Against Climate Catastrophe.* Victoria, BC: Queen's Printer Publishing.

Skundina, A. M., Kulovaa, T. L., and Yaroslavtsev, A. B. (2018). Sodium-Ion Batteries (a Review). *Russian Journal of Electrochemistry* 54, 113–152.

Solovyov, M., Pardo, E., Šouc, J., Gömöry, F., Skarba, M., Konopka, P., Pekaríčková, M., and Janovec, J. (2013). Non-uniformity of coated conductor tapes. *Superconductor Science and Technology* 26, 115013. 54

Spataru, C. and Bouffaron, P. (2016). Off-grid energy storage. In Letcher, T. M., ed., *Storing Energy - with Special Reference to Renewable Energy Sources,* Amsterdam: Elsevier. Chapter 22 pp. 477-497. DOI: 10.1016/B978-0-12-803440-8.00022-1.

Thomas, G. and Parks, G. (2006). *Potential Roles of Ammonia in a Hydrogen Economy: A Study of Issues Related to the Use Ammonia for On-Board Vehicular Hydrogen Storage.* U.S. Department of Energy. Available at https://www.energy.gov/sites/prod/files/2015/01/f19/fcto_nh3_h2_storage_white_paper_2006.pdf. 92, 93

Tinkham, M. (2004). *Introduction to Superconductivity*, 2nd ed. New York: Dover.

Tixador, P. (2008). Superconducting magnetic energy storage: Status and perspective. *IEEE/CSC & ESAS European Superconductivity News Forum*, No. 3 (January 2008). Available at http://snf.ieeecsc.org/sites/ieeecsc.org/files/CR5_Final3_012008.pdf. 55

Uchida, S. (2015). *High Temperature Superconductivity: The Road to Higher Critical Temperature*. Tokyo: Springer Japan.

Vancoillie, J., Demuynck, J., Sileghem, L., Van De Ginste, M., Verhelst, S., Brabant, L., and Van Hoorebeke, L. (2013). The potential of methanol as a fuel for flex-fuel and dedicated spark-ignition engines. *Applied Energy* 102, 140–149. DOI: 10.1016/j.apenergy.2012.05.065. 87

Vanek, F. M., Albright, L. D., and Angenent, L. T. (2016). *Energy Systems Engineering: Evaluation and Implementation*, 3rd ed. New York: McGraw Hill.

Vetter, M. and Lux, S. (2016). Rechargeable batteries with special refereence to lithium-ion batteries. In Letcher, T. M., ed., *Storing Energy - with Special Reference to Renewable Energy Sources*. Amsterdam: Elsevier. Chapter 11, pp. 205-225. DOI: 10.1016/B978-0-12-803440-8.00011-7.

Wakihara, M. (2001). Recent developments in lithium ion batteries. *Materiala Science and Engineering* R33, 109–134). DOI: 10.1016/S0927-796X(01)00030-4. 15

Wilson, J. R. and Burgh, G. (2008). *Energizing Our Future: Rational Choices for the 21st Century*. Hoboken, NJ: Wiley.

Yabuuchi, N. and Komaba, S. (2014). Recent research progress on iron- and manganese-based positive electrode materials for rechargeable sodium batteries. *Science and Technology of Advanced Materials* 15, 043501. DOI: 10.1088/1468-6996/15/4/043501. 16

Yong, Y. L., Zhou, Q. X., Li, X. H., and Lv, S. J. (2016). The $H_{60}Si_6C_{54}$ heterofullerene as high-capacity hydrogen storage medium. *AIP Advances* 6, 075321.

Author's Biography

Richard A. Dunlap received a B.S. in Physics from Worcester Polytechnic Institute in 1974, an A.M. in Physics from Dartmouth College in 1976, and a Ph.D. in Physics from Clark University in 1981. Since receiving his Ph.D., he has been on the Faculty at Dalhousie University. He was appointed Faculty of Science Killam Research Professor in Physics from 2001–2006 and served as Director of the Dalhousie University Institute for Research in Materials from 2009–2015. He currently holds an appointment as Research Professor in the Department of Physics and Atmospheric Science. Prof. Dunlap has published more than 300 refereed research papers and his research interests have included, magnetic materials, amorphous alloys, critical phenomena, hydrogen storage, quasicrystals, superconductivity, and materials for advanced batteries. Much of his work involves the application of nuclear spectroscopic techniques to the investigation of solid-state properties. He is the author of seven previous books: *Experimental Physics: Modern Methods* (Oxford, 1988); *The Golden Ratio and Fibonacci Numbers* (World Scientific, 1997); *An Introduction to the Physics of Nuclei and Particles* (Brooks/Cole, 2004); *Sustainable Energy* (Cengage, 1st ed. 2015, 2nd ed., 2019); *Novel Microstructures for Solids* (Morgan & Claypool, 2018); *Particle Physics* (Morgan & Claypool, 2018); and *The Mössbauer Effect* (Morgan & Claypool, 2019).

Printed in the United States
by Baker & Taylor Publisher Services